Treatability of Perchlorate in Groundwater Using Ion Exchange Technology—Phase II

The mission of the Awwa Research Foundation (AwwaRF) is to advance the science of water to improve the quality of life. Funded primarily through annual subscription payments from over 1,000 utilities, consulting firms, and manufacturers in North America and abroad, AwwaRF sponsors research on all aspects of drinking water, including supply and resources, treatment, monitoring and analysis, distribution, management, and health effects.

From its headquarters in Denver, Colorado, the AwwaRF staff directs and supports the efforts of over 800 volunteers, who are the heart of the research program. These volunteers, serving on various boards and committees, use their expertise to select and monitor research studies to benefit the entire drinking water community.

Research findings are disseminated through a number of technology transfer activities, including research reports, conferences, videotape summaries, and periodicals.

Treatability of Perchlorate in Groundwater Using Ion Exchange Technology—Phase II

Prepared by:
Lee Aldridge, **Thomas Gillogly**, and **Geno Lehman**
MWH
300 North Lake, Suite 1200, Pasadena, CA 91101

and

Dennis Clifford, **Deborah Roberts**, **Xiaohua Lin**
University of Houston
Department of Civil and Environmental Engineering, Houston, TX 77204-4791

Jointly sponsored by:
Awwa Research Foundation
6666 West Quincy Avenue, Denver, CO 80235-3098

U.S. Environmental Protection Agency
1200 Pennsylvania Avenue N.W., Washington, D.C. 20460
and

East Valley Water District
1155 Del Rosa Avenue, San Bernardino CA 92413

Published by:

Awwa Research Foundation
Advancing the Science of Water®

IWA Publishing

DISCLAIMER

This study was funded by the Awwa Research Foundation (AwwaRF). AwwaRF assumes no responsibility for the content of the research study reported in this publication or for the opinions or statements of fact expressed in the report. The mention of trade names for commercial products does not represent or imply the approval or endorsement of AwwaRF. This report is presented solely for informational purposes.

Published by IWA Publishing, Alliance House, 12 Caxton Street, London SW1H 0QS, UK

Telephone: +44 (0) 20 7654 5500; Fax: +44 (0) 20 7654 5555; Email: publications@iwap.co.uk;
Web: **www.iwapublishing.com**

AwwaRF report number 91016F. Originally published by AwwaRF for its subscribers in 2004.
IWA Publishing version published 2005.

ISBN 1 84339 922 9

CONTENTS

LIST OF TABLES ix

LIST OF FIGURES xi

FOREWORD xv

ACKNOWLEDGMENTS xvii

EXECUTIVE SUMMARY xix

CHAPTER 1: INTRODUCTION 1
- Primary Project Objectives 1
- Secondary Project Objectives 3

CHAPTER 2: MATERIALS AND METHODS 5
- Pilot Plant 5
  - Process Description 5
  - Process Sampling 5
  - Operational and Design Parameters 5
  - Influent Water Quality 7
- Analytical Methods 7
  - Ion Chromatography 11
  - Quality Assurance/Quality Control 11

CHAPTER 3: INITIAL RESIN EVALUATION 13
- Determination of Operational Parameters 13
  - Polyacrylic Resin 13
  - Polystyrene Resin 15
- Resin Evaluation Summary 17
- Baseline Performance 18
  - Polyacrylic 6 Percent Sodium Chloride Performance 18
  - Polystyrene 6 Percent Sodium Chloride Performance 21
  - Polyacrylic 3 Percent Sodium Chloride Performance 23
  - Heated Regeneration 25

CHAPTER 4: BRINE TREATMENT AND REUSE 29
- Biological Brine Treatment Process 30
- Chemical Brine Treatment Process 32
- Electrolytic Brine Treatment Process 33
- Brine Treatment and Reuse Testing Objectives 33

CHAPTER 5: BIOLOGICAL BRINE TREATMENT ... 35
Conversion From Synthetic Medium Feed to Brine Feed ... 35
Response of 3 Percent NaCl Culture to Brine Feed ... 35
Response of 6 Percent NaCl Culture to Brine Feed ... 38
Investigation of Brine Effect ... 39
Nutrient Requirements ... 44
Phosphate ... 44
Mixed Cations ... 46
Acetate ... 51
Impacts on Culture Performance ... 52
Oxygen ... 52
pH ... 53
Pilot-Scale Culture Development ... 54
Pilot Inoculum ... 54
Culture Growth ... 56
Pilot Performance ... 56
Pilot Brine Treatment and Recycling ... 58
Water Quality ... 61
Operation and Maintenance ... 64
Microbiological Analysis of Cultures ... 64
Enumeration of Organisms ... 64
Results of Colony Purification and Identification ... 66
Gram Stain Results and Pure Culture Characteristics Summary ... 69
Biological Brine Treatment Conclusions ... 72

CHAPTER 6: CHEMICAL BRINE TREATMENT ... 75
PNDM Pilot Testing ... 75
Water Quality ... 78
Operation and Maintenance ... 80
PNDM Conclusions ... 81

CHAPTER 7: ELECTROLYTIC BRINE TREATMENT ... 83
Electrolytic Bench-Scale Investigation ... 83
Impact of Perchlorate on Nitrate Reduction ... 83
Dependence of Nitrate on Perchlorate Reduction ... 84
Electrolytic Pilot-Scale Evaluation ... 84
Initial Cell Conditioning and Activation ... 86
Evaluation of (Per)chlorate Reduction at Low Nitrate Conditions ... 87
Electrolytic Reduction Conclusions ... 88

CHAPTER 8: ENGINEERING ANALYSIS ... 89
Cost Model ... 89
Alternative 1 – No Brine Treatment ... 89
Alternative 2 – Brine Treatment and Reuse ... 89
Alternative 3 – Brine Treatment and Disposal ... 90
Summary of Alternatives ... 90

Cost Model Calculations 92
Capital Cost 92
Operation & Maintenance Costs 93
Engineering Assumptions for Cost Model 96
Perchlorate Removal Facilities Comparison 98
La Puente Valley County Water District Ion Exchange Facility 98
Summary 101

CHAPTER 9: PROJECT CONCLUSIONS 103

REFERENCES 107

ABBREVIATIONS 109

TABLES

2.1 Ion-exchange system design and operational parameters 8

2.2 Influent water quality summary 8

2.3 Summary of approved or standard analytical procedures 10

3.1 Summary of operating conditions selected from resin evaluation 17

3.2 Summary of perchlorate recovery during baseline evaluation (polyacrylic resin) 21

3.3 Summary of perchlorate recovery during baseline evaluation (polystyrene resin).... 24

4.1 Perchlorate leakage estimates based on UH EMCT-Windows program 29

5.1 Constituents of synthetic medium 36

5.2 Constituents of ion-exchange spent brine (process brine) 36

5.3 Volatile suspended solids (3 percent NaCl brine culture) 38

5.4 Composition of media used in the brine effect experiment (3 percent culture)......... 41

5.5 Composition of media used for the 6 percent brine effect experiment 43

5.6 Media changes used for the cation addition experiment 46

5.7 Summary of pilot culture growth 57

5.8 Seawater versus pilot cation balance 58

5.9 Biologically treated brine water quality parameters with noticeable increase 63

5.10 Counts of specific organisms in the pilot culture samples 65

5.11 Numbers of aerobic heterotrophs in the first sample 66

5.12 Summary of Gram stain observations 71

6.1 Chemically treated brine water quality parameters with noticeable increase........... 80

7.1 Laboratory perchlorate reduction data 85

8.1 Life cycle cost for conventional ion exchange with brine management alternatives 91

8.2 2500 gpm estimated capital and construction related costs 93

8.3 Capital cost components 94

8.4 Assumptions for constructed related services 94

8.5 2500 gpm estimated annual O&M costs 95

8.6 O&M cost components 95

8.7 Summary of engineering assumptions used in the ion exchange cost analysis 96

8.8 Summary of assumptions used in cost model brine treatment options 97

8.9 Perchlorate removal cost comparison 98

8.10 LPVCWD (2500 gpm) estimated annual O&M costs 100

8.11 LPVCWD (2500 gpm) estimated capital and construction related costs 100

FIGURES

1.1 Known perchlorate releases (April 2003; USEPA 2003) 2

2.1 Pilot plant trailer and influent reservoir 6

2.2 Ion-exchange pilot plant 6

2.3 Ion-exchange pilot plant process schematic 7

2.4 General influent water quality (alkalinity, temperature, pH) 9

2.5 Selected influent water quality parameters 9

2.6 On-site ion chromatograph 11

2.7 Summary of pilot testing QA/QC 12

3.1 Exhaustion profile for the initial evaluation of the polyacrylic resin 14

3.2 Polyacrylic resin regeneration evaluation (62 lbs NaCl/ft^3 resin) 14

3.3 Exhaustion profile for the initial evaluation of the polystyrene resin 15

3.4 Polystyrene resin regeneration evaluation (140 BV, 530 lbs NaCl/ft^3 resin) 16

3.5 Perchlorate elution during polystyrene resin regeneration (530 lbs NaCl/ft^3 resin) 16

3.6 Comparison of exhaustion profiles during baseline testing (polyacrylic resin) 18

3.7 Nitrate exhaustion profiles during baseline testing (polyacrylic resin) 19

3.8 Regeneration profiles during baseline testing (polyacrylic resin) 20

3.9 Detailed regeneration profiles during baseline testing (polyacrylic resin) 20

3.10 Comparison of exhaustion profiles during baseline testing (polystyrene resin) 22

3.11 Comparison of regeneration profiles during baseline testing (polystyrene resin) 22

3.12 Detailed regeneration profiles during baseline testing (polystyrene resin) 23

3.13 Comparison of exhaustion profiles during 3 percent brine baseline testing (polyacrylic resin) 24

3.14 Comparison of regeneration profiles during 3 percent brine baseline testing (polyacrylic resin) 25

3.15 Heated regeneration profiles of perchlorate and sulfate (polystyrene resin) 26

3.16 Heated regeneration profiles of bicarbonate and nitrate (polystyrene resin) 27

3.17 Composite brine anions for heated brine regeneration (polystyrene resin) 27

3.18 Comparison of bed volumes required for complete displacement of nitrate and sulfate during heated regeneration 28

4.1 Brine treatment and reuse flow schematic 30

4.2 Biological brine treatment system 31

4.3 Chemical brine treatment system 32

4.4 Electrolytic brine treatment cell 33

5.1 Perchlorate reduction in the 3 percent NaCl brine culture 37

5.2 Response of 6 percent NaCl culture to sequencing batch reactor conditions 39

5.3 Effect of brine addition on perchlorate removal by the 6 percent NaCl culture 40

5.4 Response of 6 percent NaCl culture to additional brine replacement 40

5.5 Effect of brine on perchlorate reduction: 3 percent culture 1st transfer 42

5.6 Effect of brine on perchlorate degradation: 3 percent culture 3rd transfer 42

5.7 Brine effect on perchlorate degradation: 6 percent culture 1st transfer 44

5.8 Brine effect on perchlorate degradation: 6 percent culture 3rd transfer 45

5.9 Effect of not including phosphate in the 6 percent synthetic medium on perchlorate removal 45

5.10 Effect of cations on perchlorate degradation: 3 percent culture 2nd feed 47

5.11 Effect of cations on nitrate degradation: 3 percent culture 1st feed 47

5.12 Perchlorate removal in media deficient of individual cations: 6 percent culture 2nd transfer 48

5.13 The effect of increasing concentrations of Mg^{2+} on perchlorate degradation: 3 percent NaCl 1st feed 49

5.14 The effect of increasing concentrations of Mg^{2+} on perchlorate degradation: 3 percent NaCl 3rd feed 50

5.15 The effect of increasing concentrations of Mg^{2+} on nitrate degradation: 3 percent NaCl 1st feed 50

5.16 The effect of increasing concentrations of Mg^{2+} on nitrate degradation: 3 percent NaCl 3rd feed 51

5.17 Effect of acetate on perchlorate reduction 52

5.18 Impact of sulfide and oxygen exposure on perchlorate reduction 53

5.19 Effect of pH on rate of perchlorate degradation by the 3 percent culture 54

5.20 Perchlorate reduction in cultures initiated at different initial pH values 55

5.21 Growth of 3 percent synthetic seawater culture for pilot-scale use (3 to 16 L) 56

5.22 Conversion of pilot-scale culture to brine feed (3 percent NaCl) 57

5.23 Effect of divalent cation addition on perchlorate reduction 59

5.24 Comparison of exhaustion profiles during biological brine treatment (polyacrylic resin) 59

5.25 Effect of biological brine treatment and recycle on resin performance (polyacrylic resin) 60

5.26 Composite recycled brine anions prior to biological brine treatment (polyacrylic resin) 62

5.27 Composite recycled brine anions prior to biological brine treatment (polyacrylic resin) 62

5.28 Profile of acetate through the treatment process 63

5.29 Gram stains of colonies A-1, 2, 3, B, C, and D 67

5.30 Comparison of partial 16s rRNA sequences for colony A (A-1 and A-3 were identical) and two *Pseudomonas* species 68

5.31 Photomicrograph of Gram stained sample of the pilot culture (oil immersion 1000× magnification) 70

6.1 Comparison of exhaustion profiles during chemical brine treatment (polyacrylic resin) 76

6.2 Effect of chemical brine treatment and recycle on resin performance (polyacrylic resin) 76

6.3 Composite recycled brine anions prior to chemical brine treatment (polyacrylic resin) 77

6.4 Detailed regeneration profiles during chemical brine treatment (polyacrylic resin) 78

6.5 Profile of ammonia (redundant) through the treatment process 79

6.6 Arsenic removal during chemical brine treatment 80

6.7 Hardness levels during chemical brine treatment 81

7.1 Perchlorate reduction at different scan rates (scan rate bracketed in legend) 85

7.2 Overview of cell conditioning and activation 86

7.3 Chlorate reduction at low nitrate concentrations 87

7.4 Perchlorate reduction at low nitrate concentrations 88

8.1 Schematic of ion exchange with brine treatment and reuse options 90

8.2 Life cycle cost for 2500 gpm conventional ion exchange treatment plant with brine management alternatives 91

8.3 Full-scale cost comparison of ion exchange facilities 99

FOREWORD

The Awwa Research Foundation is a nonprofit corporation that is dedicated to the implementation of a research effort to help utilities respond to regulatory requirements and traditional high-priority concerns of the industry. The research agenda is developed through a process of consultation with subscribers and drinking water professionals. Under the umbrella of a Strategic Research Plan, the Research Advisory Council prioritizes the suggested projects based upon current and future needs, applicability, and past work; the recommendations are forwarded to the Board of Trustees for final selection. The foundation also sponsors research projects through an unsolicited proposal process; the Collaborative Research, Research Applications, and Tailored Collaboration programs; and various joint research efforts with organizations such as the U.S. Environmental Protection Agency, the U.S. Bureau of Reclamation, and the Association of California Water Agencies.

This publication is a result of one of these sponsored studies, and it is hoped that its findings will be applied in communities throughout the world. The following report serves not only as a means of communicating the results of the water industry's centralized research program but also as a tool to enlist the further support of the nonmember utilities and individuals.

Projects are managed closely from their inception to the final report by the foundation's staff and large cadre of volunteers who willingly contribute their time and expertise. The foundation serves a planning and management function and awards contracts to other institutions such as water utilities, universities, and engineering firms. The funding for this research effort comes primarily from the Subscription Program, through which water utilities subscribe to the research program and make an annual payment proportionate to the volume of water they deliver and consultants and manufacturers subscribe based on their annual billings. The program offers a cost-effective and fair method for funding research in the public interest.

A broad spectrum of water supply issues is addressed by the foundation's research agenda: resources, treatment and operations, distribution and storage, water quality and analysis, toxicology, economics, and management. The ultimate purpose of the coordinated effort is to assist water suppliers to provide the highest possible quality of water economically and reliably. The true benefits are realized when the results are implemented at the utility level. The foundation's trustees are pleased to offer this publication as a contribution toward that end.

Walter J. Bishop
Chair, Board of Trustees
Awwa Research Foundation

James F. Manwaring, P.E.
Executive Director
Awwa Research Foundation

ACKNOWLEDGMENTS

This project team is indebted to the East Valley Municipal Water District for initiating the process to secure federal funding for these projects and for providing general project oversight.

The authors would like to acknowledge the financial and administrative support that has been provided by the U.S. Environmental Protection Agency (USEPA). It should be noted that the information presented herein is based on the views of the project team and not of Awwa Research Foundation (AwwaRF) or the USEPA.

The authors wish to thank Frank Blaha, project manager, of AwwaRF and the following members of the Project Advisory Committee: Gerald Guter, BasinWater, Rancho Cucamonga, Calif.; and Jerry Lowry, Lowry Engineering, Blue Hill, Maine, for their guidance and suggestions.

The authors of this report are indebted to the La Puente Valley County Water District, Calif., for its involvement in and hosting of the pilot testing.

The authors would also like to thank Calgon Carbon Corporation (Pittsburgh, Pa.) for their continued support in this project by providing a perchlorate and nitrate destruction module for pilot testing. Additionally, the authors would like to acknowledge Ionex Ltd. (Nailsworth, U.K.) for supplying all the necessary equipment and performing the bench- and pilot-scale work evaluating electrolytic brine treatment for nitrate and perchlorate destruction.

Finally, the authors would like to acknowledge the work of Darren Giles, Brian Gallagher, and Eric Bruce, and Kristie Witter from the MWH Applied Research Department. The insight and comments from Edward Bouwer, Johns Hopkins University, Baltimore, Md.; and Rhodes Trussell, MWH, Pasadena, Calif., are also appreciated. In addition to the above, quality assurance was provided by Joan Oppenheimer, MWH, Pasadena, Calif.

EXECUTIVE SUMMARY

Since the detection of perchlorate in some California groundwaters in early 1997, several studies have evaluated multiple treatment technologies for its removal. While several types of treatment systems have been designed to reduce perchlorate concentrations in contaminated water, biological treatment and ion exchange systems are among the technologies most frequently investigated. Though feasible, biological treatment of perchlorate-contaminated groundwater for potable use must first overcome significant public apprehension, even open hostility, before it can be widely applied in this manner. In the interim, the implementation of ion exchange, a proven and publicly accepted technology, is available to meet the existing treatment goals. The primary concern surrounding the use of conventional ion exchange, however, is the specialized disposal and/or treatment of the perchlorate-laden saline waste stream generated as a by-product of the process.

In this research, the long-term performance of ion exchange using disparate alternatives for brine treatment and reuse was investigated. The first phase of the project (completed 2001), focused on the feasibility of ion exchange for perchlorate removal and the development of a salt-tolerant perchlorate-reducing culture.

Expanding on Phase I of the project, Phase II focused on:

- Long-Term Performance of Conventional Ion-Exchange for Perchlorate Removal. Two different types of resin (polyacrylic and polystyrene) were selected for further testing using contaminated groundwater from a currently operating drinking water well. Operational parameters were determined and baseline performance was established for both resin types.
- Effect of Heating on the Efficiency of Regeneration for Polystyrene Resins. The efficiency of non-heated regenerant (20°C) was compared to that of a regenerant heated up to 60°C in terms of effluent water quality and perchlorate leakage control.
- Brine Treatment and Reuse. Three different brine treatment alternatives were investigated: biological treatment; a physical/chemical high temperature and pressure catalytic process; and electrolytic perchlorate and nitrate reductive process.

Based on specific experiments performed in this research, the following conclusions can be made.

CONVENTIONAL ION-EXCHANGE FOR PERCHLORATE REMOVAL

- Both polyacrylic and polystyrene resins were demonstrated to be effective at removing perchlorate for extended bed volume run lengths. Perchlorate breakthrough was observed after 750 and 1500 BVs of run length for polyacrylic and polystyrene resins, respectively. The breakthrough of bicarbonate, nitrate, and sulfate occurred as expected before perchlorate breakthrough. However, sulfate breakthrough was not observed for the polyacrylic resin.
- For certain water qualities (i.e., with moderate to high nitrate), chromatographic peaking of nitrate may limit the operation of a single vessel system. Because of a background concentration of 6 mg-N/L nitrate in the groundwater at the pilot-test site

(La Puente Valley County Water District), nitrate concentrations in the process effluent to exceeded the California Department of Health Services (CaDHS) operational nitrate goal of 8 mg-N/L. This goal is 80 percent of the maximum contaminant level (MCL) and it was exceeded shortly after 400 and 625 BV for polyacrylic and polystyrene resins, respectively.

- Partial regeneration at 25 lb/ft^3 (6.4 BV) and 35 lb/ft^3 (9 BV) using 6 percent NaCl (1.0 N) brine were selected for counter-current regeneration of the polyacrylic and polystyrene resins. Counter-current regeneration of the polyacrylic resin, using a 6 percent NaCl (1.0 N) solution, demonstrated complete removal of bicarbonate, nitrate, and sulfate within 6 BV, and perchlorate removal after 13 BV. The polystyrene resin was regenerated for up to 140 BV (530 lb/ft^3) at which point the resin continued to displace approximately 800 mg/L of perchlorate. The chosen regeneration schemes ensured that nitrate, sulfate, and bicarbonate would be completely displaced from the ion exchange resin, and that a significant fraction of the perchlorate would be removed.
- Baseline performance of both resins (polyacrylic and polystyrene) was established without brine treatment and reuse. After 20 cycles of operation, as defined above, and regeneration with virgin brine, no change in performance, treated water quality, or spent brine characteristics was detected. These results confirmed the long-term operational reliability of both resins for perchlorate removal, without a negative impact on resin performance, and served as the baseline for which three brine treatment and recycle processes were compared.
- Heated regeneration of polystyrene resins improved perchlorate displacement efficiency. Based on Phase I bench testing results, heating of the regenerant was shown to reduce the amount of salt required for regeneration of polystyrene resins and consequently reduce episodes of leaching especially when using partial regeneration. Following the baseline pilot-testing protocols, fresh polystyrene resin was operated and regenerated at three different temperatures (20, 40, and 60°C) using 45 lbs/ft^3 salt loading rate. Perchlorate measured in the composite brine after the 60°C regeneration was nearly 3500 μg/L compared to the 2000 μg/L using 20°C. This improvement can be attributed to the altered equilibrium and improved kinetics at the elevated temperatures.

BRINE TREATMENT AND REUSE

The three brine treatment strategies investigated and pilot-tested in this research were:

- Biological: Utilization of salt-tolerant microorganisms capable of the simultaneous reduction of perchlorate and nitrate in a 0.5 N to 1.0 N NaCl brine using a sequencing batch reactor.
- Physical/Chemical: Operation of a high temperature and high pressure catalytic process for perchlorate and nitrate destruction in 1.0 N NaCl (Calgon Carbon Corp. – Perchlorate and Nitrate Destruction Module).
- Electrolytic: Application of a simple bipolar electrochemical cell to reduce perchlorate and nitrate to chloride, nitrogen gas and oxygen in 2.0 N KCl (Ionex Ltd. – IXL™ Process).

- Successful reuse of spent brine requires the reduction of perchlorate. The successful reuse of spent brine in the ion-exchange process depends on the residual perchlorate level in the recycled brine following treatment to biologically, chemically, or electrolytically destroy it. Maintaining a low level of perchlorate in the recycled brine is critical because of the resins high affinity for perchlorate. Furthermore, perchlorate is a monovalent ion, which is not subject to selectivity reversal in the presence of high ionic strength NaCl brine used for regeneration. Owing to its inherent high affinity and lack of selectivity (affinity) reversal, a significant portion of any perchlorate remaining in the spent brine will re-adsorb (exchange) onto the resin during regeneration. Some of the re-adsorbed perchlorate will then exchange for chloride and leak into the column effluent immediately from the start of the subsequent cycle.
- Model predictions indicated that the equivalent fraction of perchlorate in the early leakage following regeneration would be the same as the equivalent fraction of perchlorate in the treated brine. Using the University of Houston EMCT-Windows model and by varying the amounts of perchlorate in a 6 percent brine for the regeneration of the polyacrylic resin, it was determined that the maximum allowable concentration of perchlorate in the treated 6 percent brine was approximately 0.8 mg/L (0.4 mg/L for a 3 percent brine). Therefore, the perchlorate leakage can be quickly estimated as the product of the equivalent fraction perchlorate in the brine and the total concentration of ions in the feed water.

Biological Brine Treatment

The biological brine treatment system was a 208 L (55 gal) closed reactor with multiple taps to introduce spent brine, draw-off treated brine, or sample water quality. The system was operated as a sequencing batch reactor (SBR). Spent brine (0.5 N / 3 percent NaCl) from the regeneration of the ion exchange resin, was first introduced to the nitrate and perchlorate degrading culture in the reactor and allowed to mix. Acetic acid was supplied to this culture as an electron donor for the reduction process, as the nitrate and perchlorate were biodegraded. The mixture was then allowed to settle under quiescent conditions for 1 hour and the clarified treated brine was drawn through two 5-15 μm cartridge filters to remove non-settleable bacteria. The chloride concentration of the filtered, treated brine was then adjusted before its reuse as the regenerant solution.

- The biological culture was able to consistently reduce the perchlorate and nitrate concentrations in 3 percent NaCl brine to below the treatment goals for over 20 cycles. However, an increasing trend was observed in the concentration of nitrate ions in the finished water quality at the end of exhaustion. This clearly illustrated a slow decrease in capacity of the resin for nitrate as various anions accumulated in the recycled brine and on the resin.
- Accumulation of other non-targeted anions in the recycled brine could be controlled by several methods: (1) Periodically the recycled brine can be wasted and treatment continued with a fresh batch of brine, (2) periodically waste a small portion of the recycled brine (and subsequently amend it with virgin sweet brine) to limit the accumulation of these other anions, (3) further processing of the treated brine to

remove some of these accumulating anions (i.e. passing the treated brine through an NF membrane).

- Fluoride and uranium accumulated in the brine after each cycle, as these parameters were not removed by the biological brine treatment process. Although sulfate and bicarbonate accumulated in the treated and recycled brine without deleterious effects, uranium accumulated from 175 µg/L to 1700 µg/L over 30 cycles and may pose disposal and reuse issues.
- Mixing is a critical operational component of the system. Adequate mixing is required to ensure the suspension of biomass and promote continued biomass growth. Mixing, however, must not be so aggressive that it inhibits the growth of colonies.
- Oxygen is inhibitory to the salt-tolerant perchlorate-reducing microorganisms. Care must be taken to insure that an anoxic environment is maintained during treatment of perchlorate-laden brine and that dissolved oxygen poisoning is minimized, especially, during brine replacements during and after resin regeneration. The anoxic environment can be maintained by using a nitrogen sparge and/or the addition of sufficient sulfide, which promotes a reducing environment for the microorgansims.
- Pure cultures developed on solid media were all facultative, aerobic, nitrate-reducing organisms. A microecology study performed on a sample of pilot plant culture after 20 cycles of brine regeneration revealed that the culture was a mixed culture. None were identified as known human pathogens. The one organism consistently seen in Gram stains, but resisted aerobic or anaerobic solid agar culture techniques, was a Gram negative vibrio and was suspected to be the salt-tolerant perchlorate-reducing organism.
- The balance between the monovalent and divalent cations is necessary for rapid biological ClO_4^- reduction (patent pending).

Chemical Brine Treatment

The chemical brine treatment system (PNDM – Perchlorate Nitrate Destruction Module) provided by Calgon Carbon Corporation (Pittsburgh, PA), employed a high-pressure and high-temperature catalytic process to reduce the nitrate and perchlorate in the spent brine. Following the PNDM, the treated brine was ready for reuse without additional treatment. In addition to high temperature and pressure, the PNDM employed a chemical reductant (ammonia) dose based on the measured concentrations of nitrate and perchlorate in the spent brine.

- The PNDM consistently reduced the perchlorate and nitrate concentrations in the spent brine to below detection limits (71 µg/L and 0.5 mg-N/L, respectively) for 30 recycles. Perchlorate and nitrate concentrations in the spent brine (prior to treatment) remained at approximately 3100 µg/L and 270 mg-N/L, respectively. The other major anions (sulfate and bicarbonate) were not removed by the PNDM and therefore, consistently increased.
- An increasing trend was observed in the concentration of nitrate ions in the finished water as the ion exchange bed approached exhaustion. This increase indicated a slow decrease in capacity of the resin for nitrate as various anions accumulated in the recycled brine and on the resin. The concentration of these anions could also be controlled by the same methods described in the biological brine treatment process.

- Barium, fluoride, silica, and uranium accumulated in the brine after each cycle, as these parameters were not removed by the PNDM. Although sulfate and bicarbonate also accumulated in the treated and recycled brine without deleterious effects, uranium accumulated from 175 µg/L to 1700 µg/L over 30 cycles and may pose disposal and reuse issues.
- The build-up of scale was observed in the heat exchanger following the PNDM reactor even when low hardness NaCl was used. The potential for precipitation is high under the operating conditions of the PNDM, which can lead to process failures due to the clogging of system lines.
- Arsenic adsorbed by the resin during exhaustion and displaced during regeneration, was not detected in the brine after being processed by the PNDM. The concurrent loss of hardness through the brine treatment process indicated that arsenic may have been co-precipitated with $CaCO_3$ in the heat exchanger.

Electrolytic Brine Treatment

The electrolytic brine treatment was tested to evaluate if an electrochemical cell, configured for nitrate removal, could reduce perchlorate in spent ion exchange brine.

- Initial bench-scale investigations confirmed that while nitrate reduction was not inhibited in the presence of perchlorate, perchlorate reduction would only occur if nitrate levels were below 2,000 mg/L.
- A pilot-scale IXL™ cell was initially operated to activate and condition new electrodes, however, chlorate production was observed in the presence of high nitrate (4,000 to 20,000 mg/L).
- Additional tests demonstrated that chlorate production could be minimized in the presence of low nitrate (less than 2,000 mg/L), however, perchlorate removals were limited to less than 10 percent.
- Given the results of the electrolytic pilot-testing, further development of the electrolytic process is required before this could be viewed as a viable brine treatment method for perchlorate contaminated groundwaters.

COST ANALYSIS

An engineering analysis was performed to evaluate the cost of brine treatment and reuse. Based on a cost model developed to determine the capital, operation, and maintenance costs associated with removing perchlorate from groundwater using a non-proprietary ion exchange system along with one of multiple forms of management, biological and PNDM brine treatment options appear economically feasible.

- A conventional ion exchange system with brine discharge was estimated to cost $1.23/1000 gallons ($402/AF).
- Of the biological brine treatment alternatives evaluated, a 6 percent NaCl biological brine treatment option with reuse was the least costly at $1.00/1000 gallons ($326/AF), followed by a 3 percent NaCl biological system at $1.19/1000 gallons ($386/AF).

- The PNDM brine treatment option with reuse was estimated to cost $1.42/1000 gallons ($463/AF).

With increasing demands on water supplies and the continued discovery of perchlorate impaired waters, a proven and cost-effective means of treating perchlorate is needed. The implementation of ion exchange, a proven and publicly accepted technology, is available to meet the existing treatment goals. However, the primary concern surrounding the use of conventional ion exchange is the specialized disposal or treatment of the concentrated-perchlorate saline waste stream generated as a part of the process. In this research, it was found that innovative brine treatment strategies (biological, physical-chemical, electrochemical) designed to treat regenerant brine generated from ion exchange processes can be used to effectively destroy perchlorate and nitrate facilitating brine reuse.

CHAPTER 1
INTRODUCTION

Since the discovery of perchlorate (ClO_4^-) contamination in a number of California groundwaters in 1997, it has been detected in many other locations across the country. The United States Environmental Protection Agency (USEPA 2003) estimates that groundwaters in at least 44 states have the potential to be contaminated with perchlorate (Logan 2001), and has confirmed perchlorate releases in 25 states (Figure 1.1), with California having the highest number of confirmed releases (USEPA 2003, Mayer 2003). In March 1998, the USEPA formally added perchlorate to the drinking water contaminant candidate list (CCL) (Perciasepe 1998). Its monitoring in drinking water supplies was mandated shortly after under the Unregulated Contaminants Monitoring Rule (UCMR) (Browner 1999).

The primary concerns over perchlorate toxicity are based on its interference with thyroid gland function, as well as the related possible carcinogenic, developmental, reproductive, and immunotoxic effects resulting from this interference. In adults, the thyroid helps to regulate metabolism. In addition to metabolism, the thyroid plays a major role in child development. Impairment of thyroid function in expectant mothers may impact the fetus and newborn and may result in changes in behavior, delayed development and decreased learning capabilities. Changes in thyroid hormone levels may also result in thyroid gland tumors (USEPA 2003).

California Department of Health Services (CaDHS) first established an action level in 1997 when, in cooperation with the Office of Environmental Health Hazard Assessment (OEHHA), it reviewed USEPA's 1992 and 1995 evaluations of perchlorate. USEPA, as part of its Superfund activities, had developed a "provisional" reference dose (RfD) for perchlorate, based on the chemical's effects on the thyroid gland. CaDHS established an 18-μg/L action level, which corresponded to the upper value of the 4- to 18-μg/L range that resulted from USEPA's provisional RfD.

USEPA released for public review and comment a revised draft toxicity assessment on perchlorate entitled, "Perchlorate Environmental Contamination: Toxicological Review and Risk Characterization," in January 2002. This report specified a revised draft RfD of 1 μg/L. Based on this information, CaDHS concluded that its perchlorate action level needed to be revised downward. Accordingly, on January 18, 2002, CaDHS reduced the perchlorate action level to 4 μg/L, the lower value of the 4- to 18-μg/L range that resulted from the earlier provisional RfD. In December 2002, the California Office of Environmental Health Hazard Assessment (OEHHA) released a revised draft perchlorate Public Health Goal (PHG), proposing a concentration of 2 to 6 μg/L. OEHHA's PHG, when final, contributes to CaDHS's development of an MCL for perchlorate. Currently, CaDHS continues to utilize the 4-μg/L action level.

PRIMARY PROJECT OBJECTIVES

As a collaborative effort between MWH and the University of Houston, the treatability of perchlorate in groundwater using ion-exchange technology was investigated in Phase I of this project and demonstrated that many IX resins are capable of removing perchlorate from groundwater. The resins evaluated included: (1) polyacrylic resins (Rhom and Haas IRA-458, Purolite A-850, etc.), which have moderate selectivity for perchlorate but are easy to regenerate; (2) polystyrene resins (Sybron ASB-2, SR-7, etc.), which have high selectivity for perchlorate but are difficult

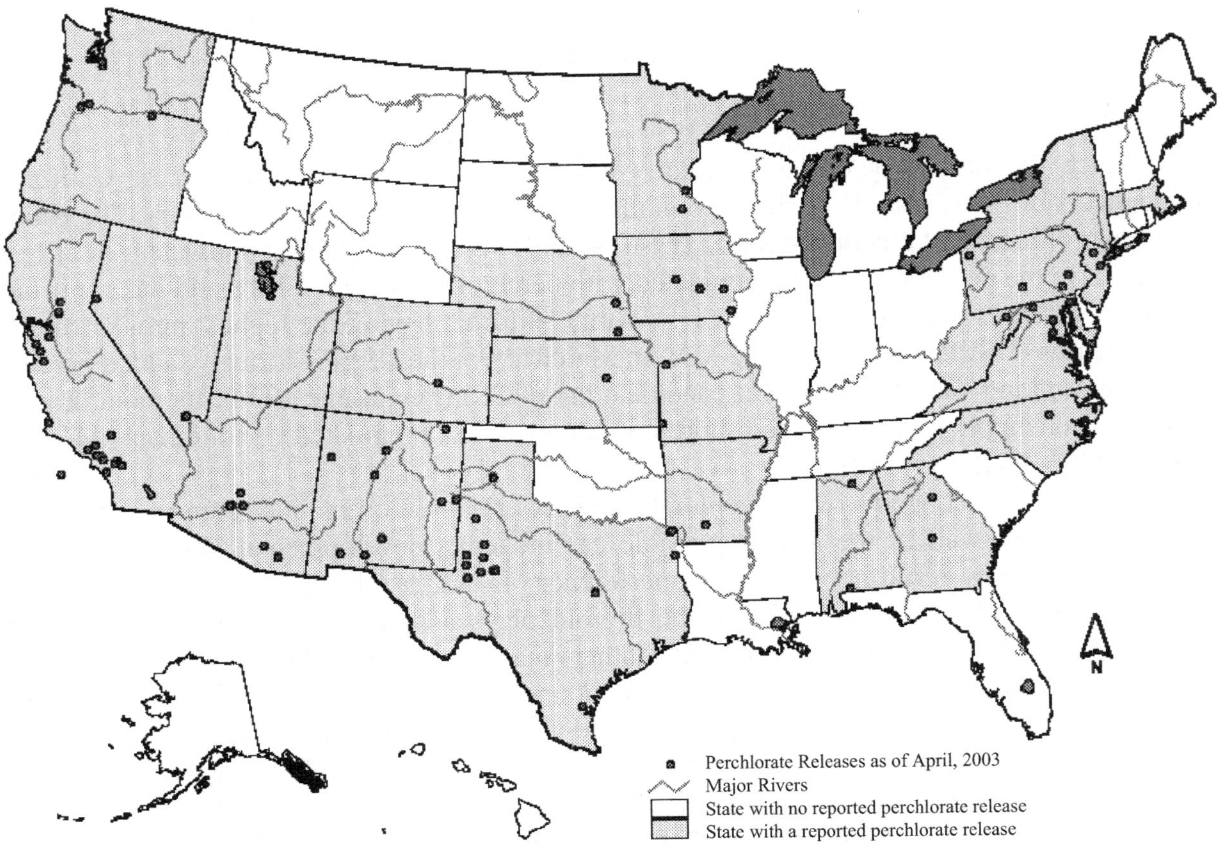

Figure 1.1 Known perchlorate releases (April 2003; USEPA 2003)

to regenerate; and (3) polyvinylpyridine resins (Reillex DP-1, HPQ, etc.), which have very high capacities for perchlorate but are commensurately difficult to regenerate. Bench-scale testing conducted during Phase I indicated that typical polyacrylic resins can process somewhere between 500 and 1000 Bed Volumes (BVs) before perchlorate breakthrough is observed (assuming 50 to 100 μg/L perchlorate, 5 mg/L nitrate, and 50 mg/L sulfate in the groundwater). To regenerate these resins, a salt loading rate of approximately 30 lbs/ft^3 would typically be required.

Expanding on Phase I of the project, Phase II focused on:

- Long-Term Performance of Conventional Ion-Exchange for Perchlorate Removal. Two different types of resin (polyacrylic and polystyrene) were selected for further pilot-testing using contaminated groundwater from a currently operating drinking water well. Operational parameters were determined and baseline performance was established for both resin types.
- Effect of Heating on the Efficiency of Regeneration for Polystyrene Resins. The efficiency of non-heated regenerant (20°C) was compared to that of a regenerant heated up to 60°C in terms of effluent water quality and leakage control.
- Brine Treatment and Reuse. Three different brine treatment alternatives were investigated: biological treatment; a physical/chemical high temperature and pressure catalytic process; and electrolytic perchlorate and nitrate reduction.

SECONDARY PROJECT OBJECTIVES

Secondary objectives that were addressed during this project included:

- Determine the post-treatment needs to make the ion-exchange effluent potable;
- Identify the by-products formed as a result of brine treatment through the physical/chemical and biological processes;
- Characterize the microbial ecology of the biological brine treatment culture;
- Evaluate the reliability of the brine treatment processes and discuss how they may be integrated into a full-scale application;
- Analyze the engineering and economic feasibility for making conventional ion exchange coupled with brine treatment, a reliable process for the removal of perchlorate from groundwater.

CHAPTER 2
MATERIALS AND METHODS

To accomplish the objectives of this research, a suite of analytical methods, experimental systems, and procedures were employed. The analytical methods were used to assess the water quality and quantify the performance of the experimental systems.

PILOT PLANT

Designed, constructed, and installed in MWH's Mobile Water Treatment Pilot Trailer during the initial periods of this project, the ion-exchange pilot plant was located in La Puente, California. This site, owned and operated by the La Puente Valley County Water District, has an active groundwater well and employs a full-scale 2,500 gpm Calgon Carbon Corporation ISEP ion-exchange system for the removal of perchlorate. A photo of the exterior of the trailer and system feed tank is shown in Figure 2.1.

Process Description

Figure 2.2 shows a picture of the pilot plant. The pilot plant included two parallel ion-exchange columns (clear PVC) that could be operated independently in either co- or counter-current exhaustion and regeneration. For the purposes of this study, the columns were operated in a counter-current mode with up-flow exhaustion and down-flow regeneration. Screens were inserted at the top and bottom of each ion-exchange column to contain the resin in the column during the exhaustion and regeneration modes. The treated water was collected in a clearwell and was also used for rinse water after regeneration. The spent brine solution was either wasted or collected in a holding tank to be fed to the brine treatment system. The treated brine was then returned to the regeneration (sweet) brine tank. A process schematic of this system is shown in Figure 2.3 to illustrate some of these details.

Process Sampling

The pilot was designed so that samples could be collected from the break tank (influent water), each column effluent during service and regeneration, clearwell, sweet brine tank, spent brine storage tank, and brine treatment system effluent. Sampling sites were selected to provide a complete analysis of a variety of processes used in this study. The frequency of sample collection was based on operational conditions and historical performance.

Operational and Design Parameters

The operational and design parameters of the system are summarized Table 2.1. It was decided that for comparative purposes, one polystyrene resin and one polyacrylic resin would be used in this study. Purolite A-850 and Bayer M500 were selected as representative polyacrylic and polystyrene resins, respectively, based on discussions with resin manufacturers and a bench-scale resin screening evaluation. During the exhaustion of these resins, a 1.7 gpm flowrate was used for each 4-inch diameter column. Given an empty bed contact time (EBCT)

Figure 2.1 Pilot plant trailer and influent reservoir

Figure 2.2 Ion-exchange pilot plant

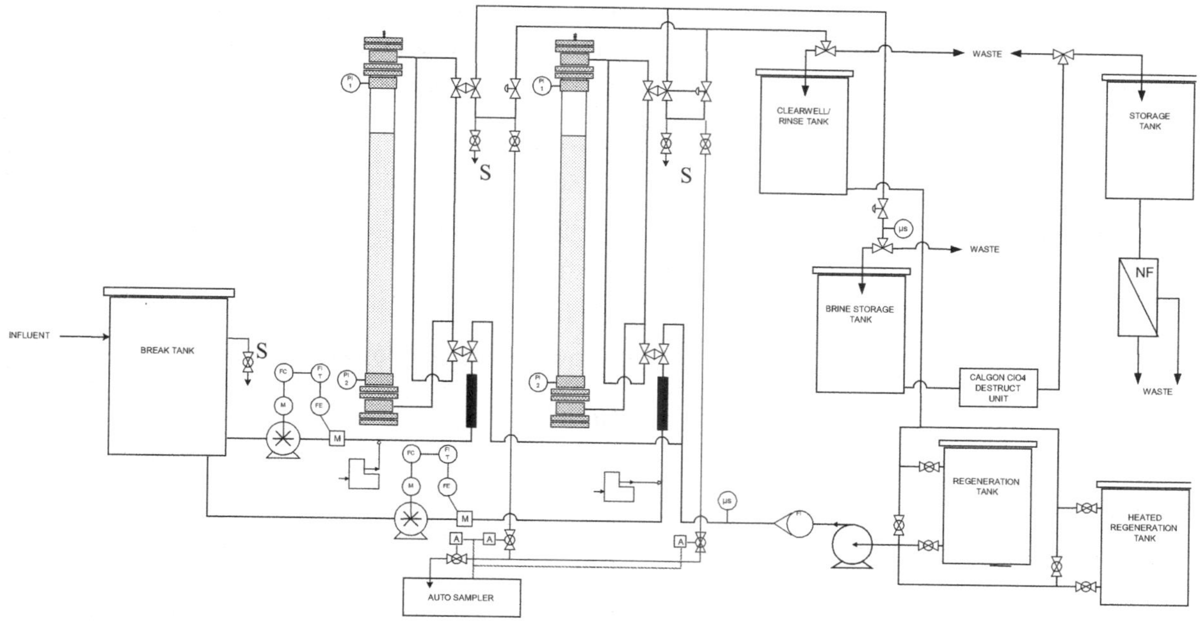

Figure 2.3 Ion-exchange pilot plant process schematic

of 1.5 minutes, this translated into an up-flow service loading rate of 4.9 gpm/ft^3 resin. A 17.4 cm/min velocity (19.5 gpm/ft^2) was used for counter-current regeneration, corresponding to a 7-min EBCT. Salt loading rates of 35 lbs/ft^3 and 25 lbs/ft^3 were used during normal operation of the polystyrene resin and the polyacrylic resin, respectively.

Influent Water Quality

Selected parameters were monitored in the influent water quality. The influent groundwater was generally stable for most of the parameters monitored, as summarized in Table 2.2. The alkalinity, temperature and pH of the water during pilot testing are shown in Figure 2.4. The concentrations of other selected water quality parameters (perchlorate, sulfate, and nitrate) are displayed in Figure 2.5. Apparent from this figure was a decreasing trend in the perchlorate concentration since the beginning of the study, most likely due to the movement of the perchlorate plume in the groundwater.

ANALYTICAL METHODS

Most of the water quality parameters were measured following Standard Methods (APHA, AWWA, and WEF 1998) or USEPA methods. A summary of analytical procedures used, both approved and other validated standard methods, is provided in Table 2.3. A discussion of modified and non-standard methods is also included.

Table 2.1
Ion-exchange system design and operational parameters

Parameter	Column 1	Column 2
	System Design	
Resin Name	A-850	M500
Column Diameter	4.0 in	4.0 in
Resin Bed Depth	4.0 ft	4.0 ft
Bed Volume	0.35 ft^3	0.35 ft^3
Flow Mode	up-flow service and down-flow regeneration	
	Exhaustion	
EBCT	1.5 min	1.5 min
Service Flowrate	1.7 gpm	1.7 gpm
Service Loading Rate	4.9 gpm/ft^3	4.9 gpm/ft^3
Surface Loading Rate	19 gpm/ft^2	19 gpm/ft^2
Run Length	400 BV*	625 BV
	Regeneration	
Regenerant	NaCl	NaCl
EBCT	7 min	7 min
Regenerant Flowrate	1416 mL/min	1416 mL/min
Downflow Velocity (Q/A)	17.4 cm/min	17.4 cm/min
Regenerant Strength	6 (3) percent	6 percent
Salt Loading Rate	25 lbs/ft^3	35 lbs/ft^3
Regeneration Period	45 min	63 min
Bed Volumes of Regenerant	6.4	9
Volume of Regenerant (total)	64 L	89 L
Bed Volumes of Rinse	3	3

* Bed volumes.

Table 2.2
Influent water quality summary

Parameter	Unit	Median	Average	Standard deviation	Range
Temperature	°C	20.6	21.7	2.86	15 - 26.7
Turbidity	NTU	0.4	0.4	0.08	0.31 - 0.43
pH		8.1	8.1	0.15	7.68 - 8.97
Total Alkalinity	mg $CaCO_3$/L	182	181	11.5	120 - 208
Conductivity	µS	491	491.0	2.8	489 - 493
Perchlorate	µg/L	55	54	6.2	40 - 71
Nitrate as N	mg/L	5.9	5.7	0.42	4.3 - 6.9
Sulfate	mg/L	30.6	31	3.7	24.8 - 39.7

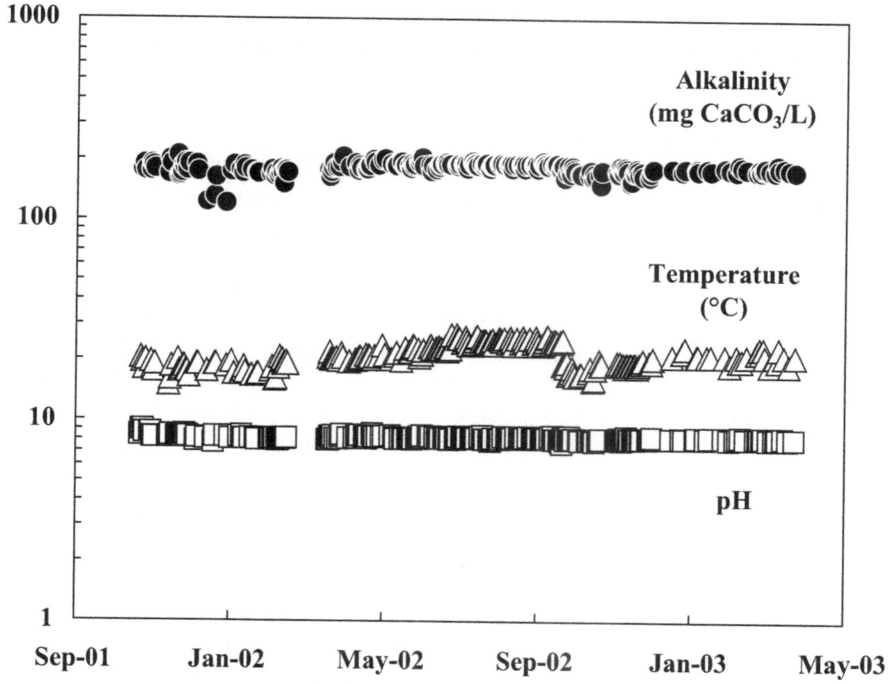

Figure 2.4 General influent water quality (alkalinity, temperature, pH)

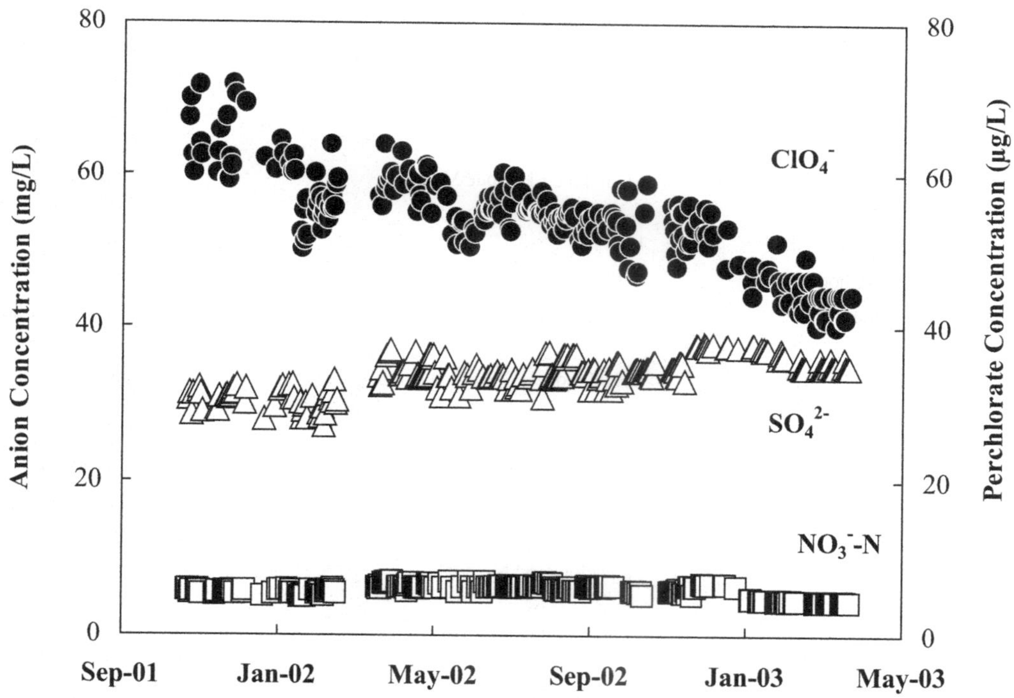

Figure 2.5 Selected influent water quality parameters

Table 2.3
Summary of approved or standard analytical procedures

Parameter	Method number	Method title	Reference
ClO_4^-	EPA Method 314.0*	Determination of Low Concentrations of Perchlorate in Drinking Water Using Ion Chromatography	USEPA
	Modified USEPA Method 300.1†	Analysis of Low Concentrations of Perchlorate in Drinking Water and Groundwater by Ion Chromatography	Dionex
ClO_3^-, ClO_2^-	Modified USEPA Method 300.1*	Determination of Inorganic Anions in Environmental Waters with a Hydroxide-selective Column	Jackson et al. 2000
	EPA Method 300.1†	Determination of Inorganic Anions in Drinking Water by Ion Chromatography	USEPA
Cl^-	Modified USEPA Method 300.1*	Determination of Inorganic Anions in Environmental Waters with a Hydroxide-selective Column	Jackson et al. 2000
	EPA Method 300.1†	Chloride-Ion Chromatography Method	USEPA
NO_3^-	Modified USEPA Method 300.1*	Determination of Inorganic Anions in Environmental Waters with a Hydroxide-selective Column	Jackson et al. 2000
	EPA Method 300.1†	Ion Chromatographic Method	USEPA
NO_2^-	Modified USEPA Method 300.1*	Determination of Inorganic Anions in Environmental Waters with a Hydroxide-selective Column	Jackson et al. 2000
	EPA Method 300.1†	Ion Chromatographic Method	USEPA
SO_4^{2-}	Modified USEPA Method 300.1*	Determination of Inorganic Anions in Environmental Waters with a Hydroxide-selective Column	Jackson et al. 2000
	EPA Method 300.1†	Ion Chromatographic Method	USEPA
H_2 (g)	Non-standard method	Orbisphere* Trace Analytical†	
Acetate	4110	Determination of Anions by Ion Chromatography	Standard Methods (APHA, AWWA, and WEF 1998)
TOC/DOC	5310C	Persulfate Ultraviolet Method	Standard Methods (APHA, AWWA, and WEF 1998)
DO	4500-O G.	Membrane Electrode Method	Standard Methods (APHA, AWWA, and WEF 1998)
Ca^{2+}	Method 200.7	ICP	USEPA
Mg^{2+}	Method 200.6	ICP	USEPA
Total Hardness	2340C	EDTA Titrimetric Method	Standard Methods (APHA, AWWA, and WEF 1998)
Bicarbonate	4500-CO_2 D.	Calculation	Standard Methods (APHA, AWWA, and WEF 1998)
NH_3	4500-NH_3 F.	Phenate Method	Standard Methods (APHA, AWWA, and WEF 1998)
Total Alkalinity	2320B	Titration Method	Standard Methods (APHA, AWWA, and WEF 1998)
Turbidity	2130B	Nephelometric Method	Standard Methods (APHA, AWWA, and WEF 1998)
Place Counts	9215B	Pour Plate Method	Standard Methods (APHA, AWWA, and WEF 1998)
Temperature	2550B	Laboratory and Field Methods	Standard Methods (APHA, AWWA, and WEF 1998)
pH	4500H^+	Electrometric Method	Standard Methods (APHA, AWWA, and WEF 1998)
Specific Conductivity	2510B	Laboratory Method	Standard Methods (APHA, AWWA, and WEF 1998)

* Pilot-scale.
† Bench-scale.

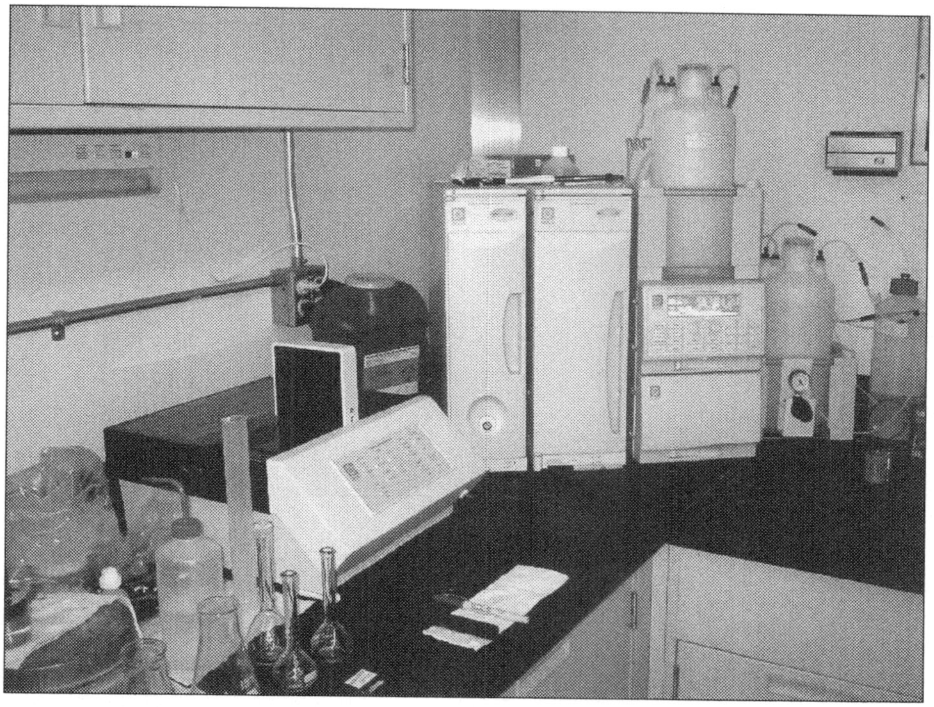

Figure 2.6 On-site ion chromatograph

Ion Chromatography

Utilizing the latest technology at pilot-scale, perchlorate was analyzed on-site by ion chromatography (IC) using a Dionex DX-320 with conductivity detection using an AS-16 column, a 1000 µL loop, an EG-40 eluent generator, and an autosampler as shown in Figure 2.6. USEPA Method 314.0 was followed as the analytical protocol. Based on seven injections of a 2-µg/L standard over the period of one week, the MDL was determined to be 0.7 µg/L. The lowest standard used during calibration was 2 µg/L.

All anions other than perchlorate (chloride, chlorate, chlorite, nitrate, nitrite, and sulfate) were analyzed on the same system using an AS-17 column, a 10-µg/L loop, and an EG-40 eluent generator configured for gradient analysis. Complete resolution of all of these anions of interest was evaluated in a single run using the EG-40 to produce a hydroxide eluent gradient based on USEPA Method 300.1 modified for use with a hydroxide-selective column (Jackson et al. 2000).

Quality Assurance/Quality Control

Quality Assurance (QA) was maintained through the continued use of the approved Quality Assurance Project Plan (QAPP). The objective of this plan was to ensure that the results and conclusions derived from the project were based on valid data of known and documented quality. The QAPP consisted of the following major elements: 1) quality control (QC), which includes sampling procedures, sample chain of custody, measurement of data, calibration procedures and frequency, and data reduction, validation, and reporting; 2) quality assessment, which includes performance and system audits, and correction action; and 3) reporting.

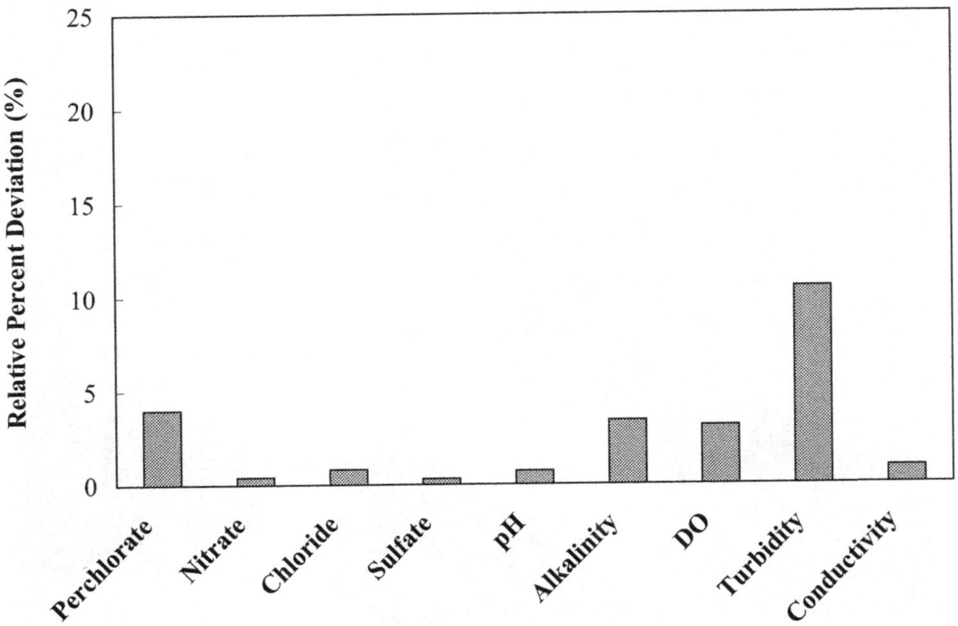

Figure 2.7 Summary of pilot testing QA/QC

Project QC was maintained to control analytical measurement errors and included the following elements: sampling procedures, sample chain of custody, development and strict adherence to principles of good laboratory practice (using qualified staff and reliable, well-maintained equipment), verified and specified analytical methods for measurements (including sensitivity, selectivity, precision, and accuracy), and consistent use of standard instrument operation procedures (including calibration and standardization procedures, minimum reporting limits-MRLs, and/or method detection limits-MDLs, and frequency). Project QC resulted in a measurement system operating in a state of statistical control whereby errors are reduced to acceptable levels.

The data management system involved the use of computer spreadsheets as well as a protocol for the entry, verification, and presentation of the data in appropriate tabular and graphical forms. Field data (lab evaluations and field sampling) was recorded by hand on specially prepared data sheets for all operational and water quality measurements and/or observations. These data were entered into the appropriate spreadsheet. Following entry, the spreadsheet was printed out and the hard copy checked against the handwritten datasheet. Any corrections were noted on the hard copy; corrected on the screen and saved followed by a print out of the updated spreadsheet. Data was exported directly from the spreadsheets for use in graphics, statistics, and water quality models.

As specified in the QAPP, the overall precision of water quality analysis performed by the project team was verified by replicate analysis and calculation of the Relative Percent Deviation (RPD) and the goal was defined to be less than or equal to 25 percent. As seen in Figure 2.7, the QA goal was consistently met for all the analyses with the majority of analysis having an RPD of less than 5 percent. Although still well below the QA goal, turbidity analysis had a RPD of 10 percent due to the extremely low levels of turbidity observed and the variability at these levels.

CHAPTER 3
INITIAL RESIN EVALUATION

Both the polyacrylic (Purolite A-850) and polystyrene (Bayer M500) resins were initially evaluated by operating them to perchlorate breakthrough in order to determine the appropriate run length to be used for baseline testing. After perchlorate breakthrough, the resins were regenerated in a counter-current mode to determine the appropriate long-term operational conditions for the regeneration of the resins.

DETERMINATION OF OPERATIONAL PARAMETERS

Polyacrylic Resin

The polyacrylic resin was exhausted for 900 BV, in which breakthrough of perchlorate, nitrate, and bicarbonate occurred. Sulfate breakthrough was not observed during this period. Figure 3.1 shows perchlorate breakthrough occurring after approximately 750 BV of water processed. Nitrate and bicarbonate were observed in the effluent much earlier in the run at approximately 275 BV and 100 BV, respectively. The anticipated decrease in the effluent chloride ion concentration can be seen during the course of the exhaustion cycle as less chloride was available to be displaced from the resin over time by additional bicarbonate, nitrate, perchlorate, and sulfate ions.

During this run, chromatographic peaking of nitrate concentrations in excess of California Department of Health Services (CaDHS) operational nitrate goal of 8 mg-N/L, 80 percent of the maximum contaminant level (MCL), was observed. As a result, the polyacrylic resin was operationally limited by nitrate in this water, unless a sufficient number of ion-exchange vessels could be operated in parallel to ensure that a blended effluent did not exceed the operational limit. It was decided that for the purposes of this study, the pilot plant should be operated such that it produces water of potable quality. Consequently, the run length selected for the baseline testing was 425 BV. Since perchlorate was not observed in the effluent until approximately 750 BV, a 425 BV run length was deemed suitable for both the removal of perchlorate and nitrate from the groundwater.

Counter-current regeneration of the polyacrylic resin was performed using a 6 percent sodium chloride solution until concentrations of perchlorate and nitrate remained below analytical detection limits. Figure 3.2 shows the profile of anions displaced during the regeneration cycle. The resin was regenerated with 16 bed volumes of regenerant solution at a salt loading rate of 62 lbs NaCl/ft^3 of resin. Nitrate, sulfate, and bicarbonate were completely removed from the resin after 6 BV of regeneration, and perchlorate was removed after 13 BV of regeneration.

Based on these results, a partial regeneration of 25 lb/ft^3 (6.4 BV) was selected for the polyacrylic resin during baseline testing. This would ensure that no nitrate, sulfate, nor bicarbonate would accumulate on the resin over time. It was anticipated to displace a large fraction of the perchlorate, especially since during regular operation (i.e., shorter run lengths) less perchlorate would accumulate on the resin.

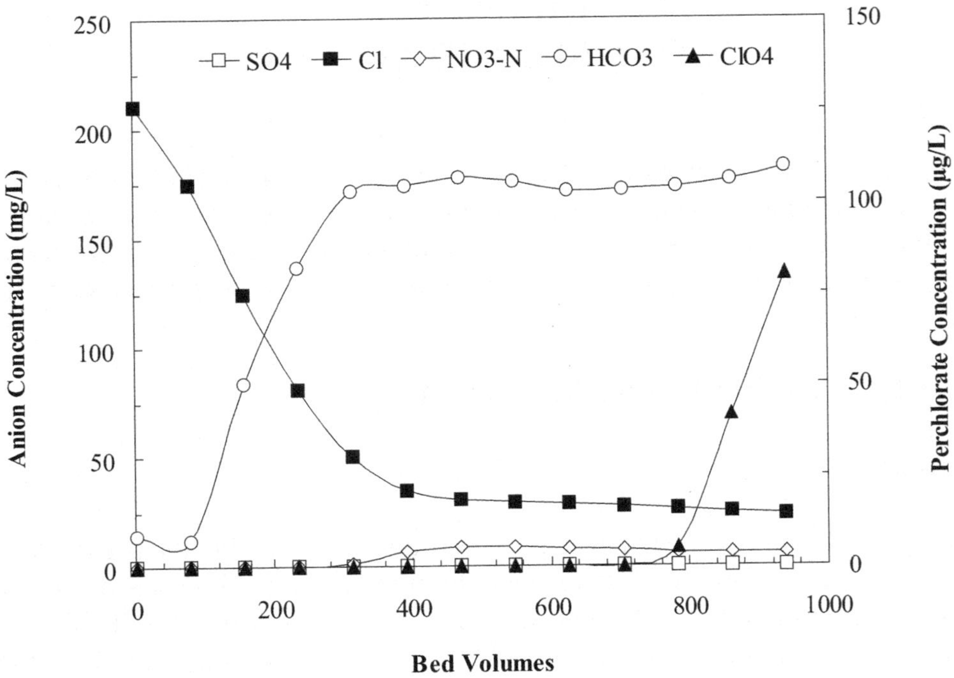

Figure 3.1 Exhaustion profile for the initial evaluation of the polyacrylic resin

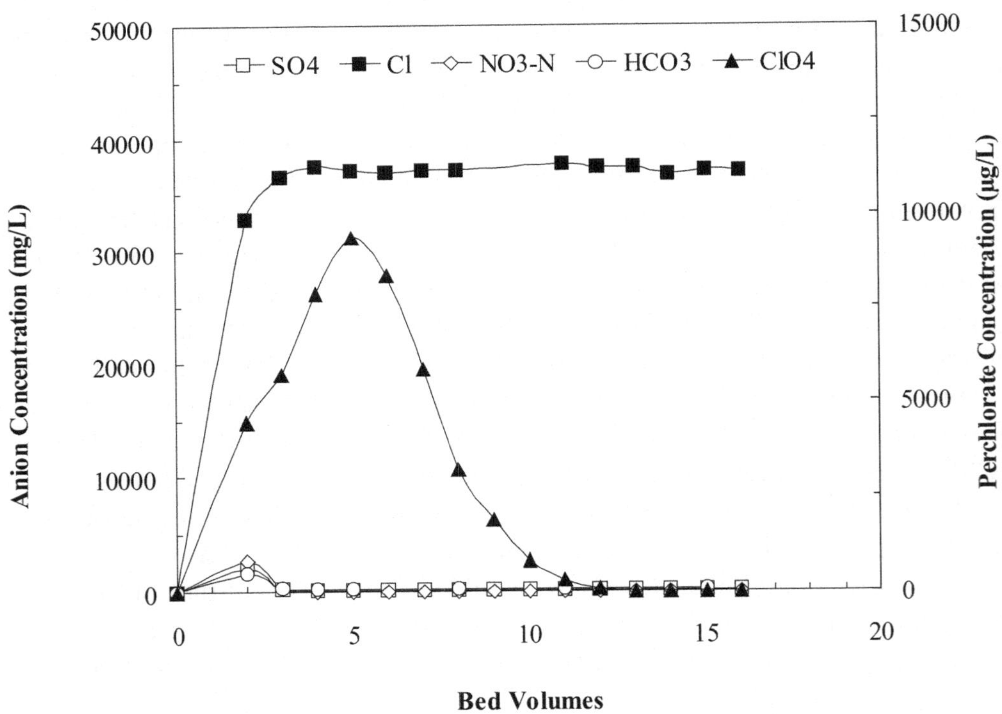

Figure 3.2 Polyacrylic resin regeneration evaluation (62 lbs NaCl/ft^3 resin)

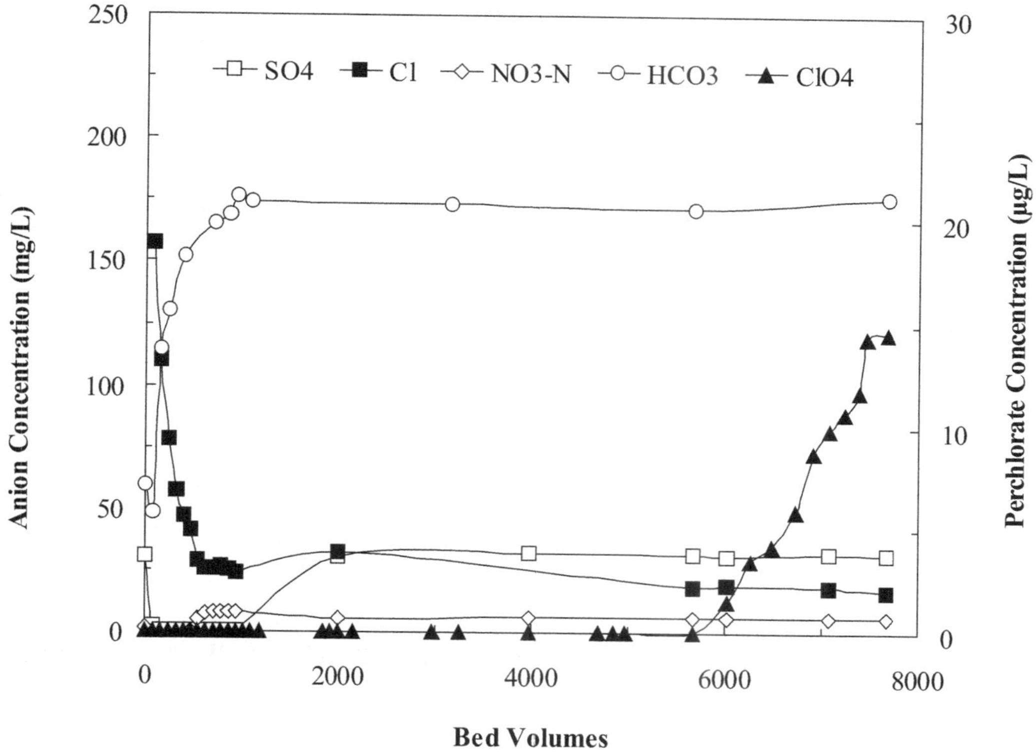

Figure 3.3 Exhaustion profile for the initial evaluation of the polystyrene resin

Polystyrene Resin

The polystyrene resin was initially operated for 7600 BV, during which time, the breakthrough of perchlorate, nitrate, sulfate and bicarbonate were all observed. Perchlorate was observed in the effluent after approximately 5700 BV, as shown in Figure 3.3. Bicarbonate, nitrate and sulfate were also detected in the effluent at approximately 100 BV, 400 BV, and 1500 BV, respectively. The characteristic decrease in the effluent chloride ion concentration corresponded predominantly to the adsorption of bicarbonate from the water.

As with the polyacrylic resin, a slight chromatographic peaking of nitrate was observed. While the nitrate peak remained below the 10 mg-N/L MCL, it exceeded the 8 mg-N/L CaDHS operational goal for a "single vessel" type of operation. As such, the polystyrene resin was also nitrate limited under these conditions. The run length selected for baseline testing was 625 BV. Since perchlorate breakthrough did not occur until approximately 6000 BV, a 625 BV run length was short enough to ensure both the removal of perchlorate and nitrate from the groundwater.

Counter-current regeneration of the polystyrene resin was performed using a 6 percent NaCl solution. Figure 3.4 shows the regeneration profile of anions adsorbed during the exhaustion cycle. The resin was regenerated with 140 BV of regenerant solution at a salt loading rate of 530 lbs NaCl/ft^3 of resin. Nitrate, sulfate, and bicarbonate were completely removed from the resin after 8 BV of regeneration. Figure 3.5 shows the regeneration profile of perchlorate. The perchlorate concentration in the regenerant brine peaked at approximately 6000 µg/L after 2 BV of regeneration. Even after 140 BV (530 NaCl/ft^3) of regeneration, however, analysis of the brine

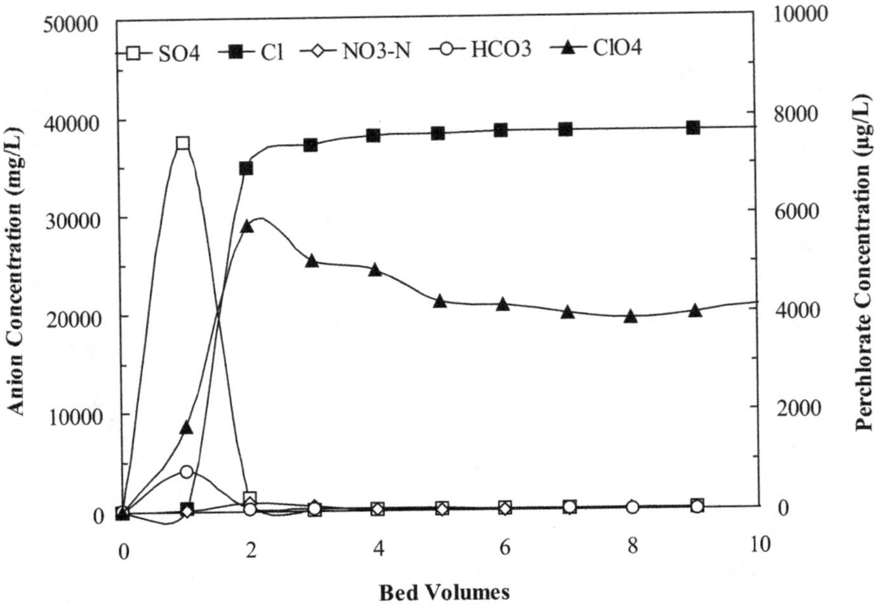

Figure 3.4 Polystyrene resin regeneration evaluation (140 BV, 530 lbs NaCl/ft^3 resin)

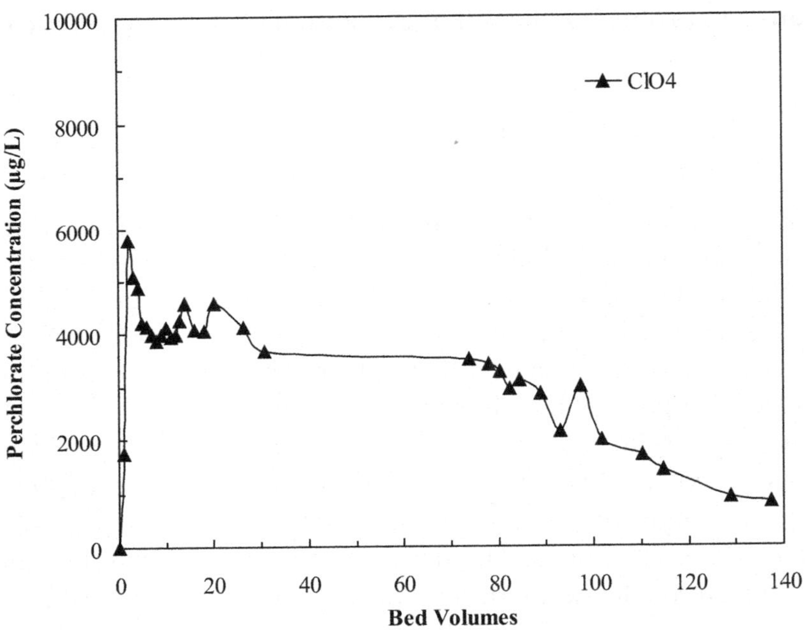

Figure 3.5 Perchlorate elution during polystyrene resin regeneration (530 lbs NaCl/ft^3 resin)

Table 3.1
Summary of operating conditions selected from resin evaluation

Resin Name	Purolite A-850	Bayer M500
Resin Matrix	Polyacrylic	Polystyrene
Resin Bed Volume	0.35 ft^3	0.35 ft^3
	Exhaustion	
Run Length	425 BV	625 BV
Flow Rate	1.7 gpm	1.7 gpm
EBCT	1.5 min	1.5 min
	Regeneration	
Operational Flow	Counter-Current	Counter-Current
Salt Loading Rate	25 lbs/ft^3	35 lbs/ft^3
Run Length	6.4 BV	9 BV
Flow Rate	0.37 gpm	0.37 gpm
EBCT	7 min	7 min
Exhaust:Regen BV Ratio	66.4	66.7

indicated that approximately 800 μg/L perchlorate was still being displaced from the ion-exchange resin. Regeneration of the polystyrene resin was terminated after 140 BV because the profile demonstrated the large quantity of salt required for the complete regeneration.

A salt loading rate of 35 lb/ft^3 (9 BV) was selected for the partial regeneration of the polystyrene resin during baseline testing. This amount of regeneration ensured that nitrate, sulfate, and bicarbonate would be completely displaced from the ion-exchange resin. The regeneration was also expected to remove a significant fraction of the sorbed perchlorate.

RESIN EVALUATION SUMMARY

Results from evaluation of both the polyacrylic and polystyrene resins indicated the total amount of water that could be processed before perchlorate was observed in the effluent of the system. The resin evaluation also illustrated the significantly higher perchlorate selectivity with the polystyrene resin as compared to the polyacrylic resin. The operation of both resins, however, was limited by the chromatographic peaking of nitrate in excess of the CaDHS operational goal. Should nitrate levels decrease in this groundwater, it may be possible to process more water during each cycle. For now, Table 3.1 summarizes the preliminary operating conditions determined for baseline testing of each resin.

It is interesting to note that while these initial tests clearly demonstrated that the polystyrene resin required larger quantities of salt for a complete regeneration than the polyacrylic resin, on a per unit perchlorate adsorbed basis, similar quantities of salt were required for regeneration.

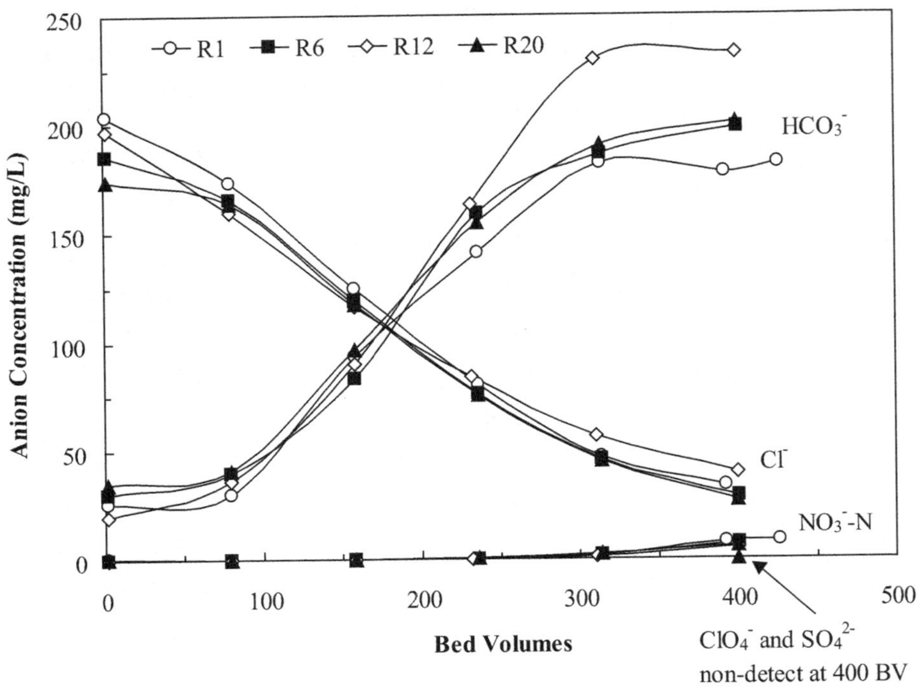

Figure 3.6 Comparison of exhaustion profiles during baseline testing (polyacrylic resin)

BASELINE PERFORMANCE

Once initial operating conditions were established, baseline performance was obtained by treating contaminated groundwater to determine the long-term performance of conventional ion-exchange technology for perchlorate removal and to serve as a reference for subsequent brine treatment and recycling experiments. Each type of resin (polyacrylic and polystyrene) was exhausted and regenerated with a 6 percent NaCl brine solution for 20 cycles. Baseline performance was also established for the polyacrylic resin regenerated with a 3 percent NaCl solution. Finally, an experiment was performed to evaluate the effect of heating the brine used for regeneration of the polystyrene resin.

Polyacrylic 6 Percent Sodium Chloride Performance

Using the operating conditions determined during the initial resin evaluation, the polyacrylic resin was exhausted for 425 BV at a flowrate of 1.7 gpm, and regenerated with 25 lbs/ft^3 using 6 percent NaCl brine. Twenty exhaustion and regeneration cycles were performed to establish the baseline performance. Figure 3.6 summarizes the exhaustion profiles of perchlorate, nitrate, sulfate, chloride, and bicarbonate observed during selected runs (Run 1, 6, 12, and 20). Effluent analyses for the other runs indicated similar performance. As seen in these anion profiles, a high degree of repeatability was observed during the course of the baseline testing. This consistency indicated that bicarbonate and nitrate were not accumulating on the resin. However, the exhaustion data alone were insufficient to determine if other anions, like perchlorate and sulfate, were accumulating on the ion-exchange resin with the given regeneration procedure.

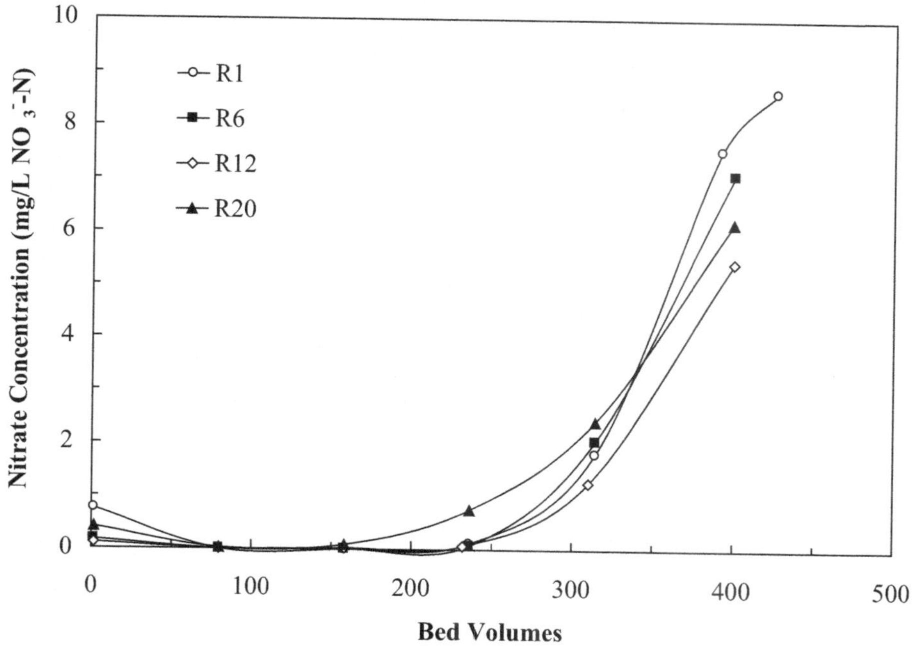

Figure 3.7 Nitrate exhaustion profiles during baseline testing (polyacrylic resin)

Closer inspection of the effluent nitrate concentration throughout the baseline testing also revealed a high level of reproducibility. As shown in Figure 3.7, during the first run of baseline testing the nitrate level measured in the column effluent at 425 BV was 8.6 mg-N/L. As this concentration was in excess of the CaDHS operational treatment goal, 80 percent of the 10 mg-N/L MCL, the subsequent ion-exchange runs were shortened to 400 BV. This resulted in a final effluent nitrate concentration consistently below the operational nitrate goal.

Partial regeneration was performed in a counter-current mode at a flow rate of 0.37 gpm for 6.4 BV and a salt loading rate of 25 lbs/ft^3. Figure 3.8 is a comparison of the regeneration profiles of perchlorate, nitrate, sulfate, and chloride observed during selected runs. Nitrate and sulfate were completely displaced from the resin after 3 BV. Measurable amounts of perchlorate were observed at the end of each of the twenty regeneration cycles, with the average concentration calculated to be 680 μg/L. The chloride concentration increased for 3 BV and then remained relatively constant for the rest of the regeneration at 3700 mg/L.

A closer look at the regeneration curves for each individual anion displaced during regeneration (Figure 3.9) indicated that the total mass of anions adsorbed by the resin during these runs varied. This variation was likely to be the result of the small but regular variability in the influent water quality concentrations of these anions (perchlorate, nitrate, sulfate, and bicarbonate). The lack of a shift in elution time or consistent increase in peak area indicated that these anions were not accumulating on the resin in significant quantities, within the twenty baseline cycles.

As measurable concentrations of perchlorate were in the brine at the end of the regeneration cycle, it was expected that after prolonged operation, the concentration of perchlorate would build up on the resin. Once this concentration accumulated to the point that perchlorate leakage would be observed, a complete regeneration of the resin with respect to perchlorate would be

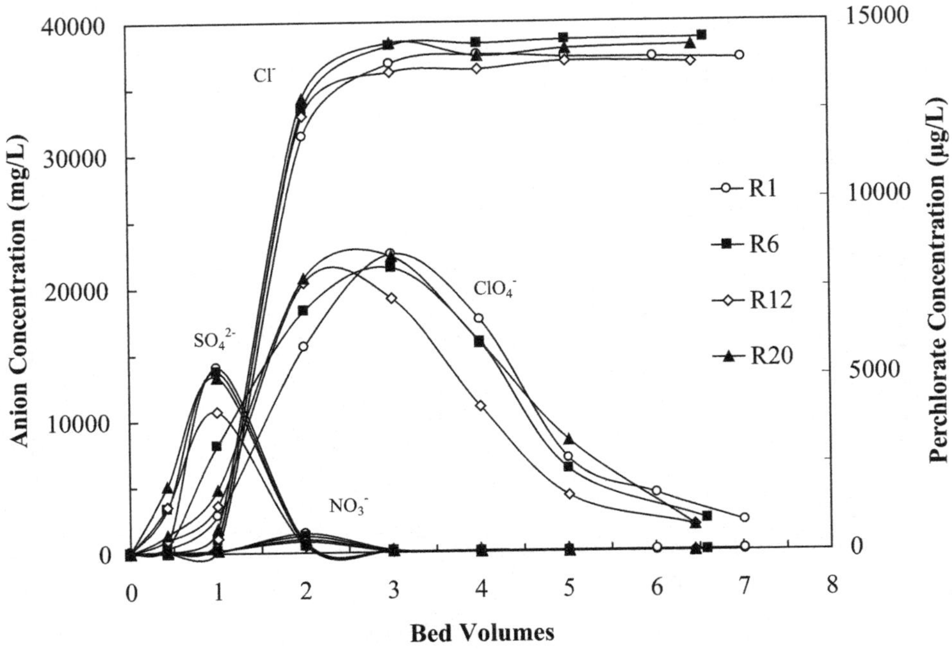

Figure 3.8 Regeneration profiles during baseline testing (polyacrylic resin)

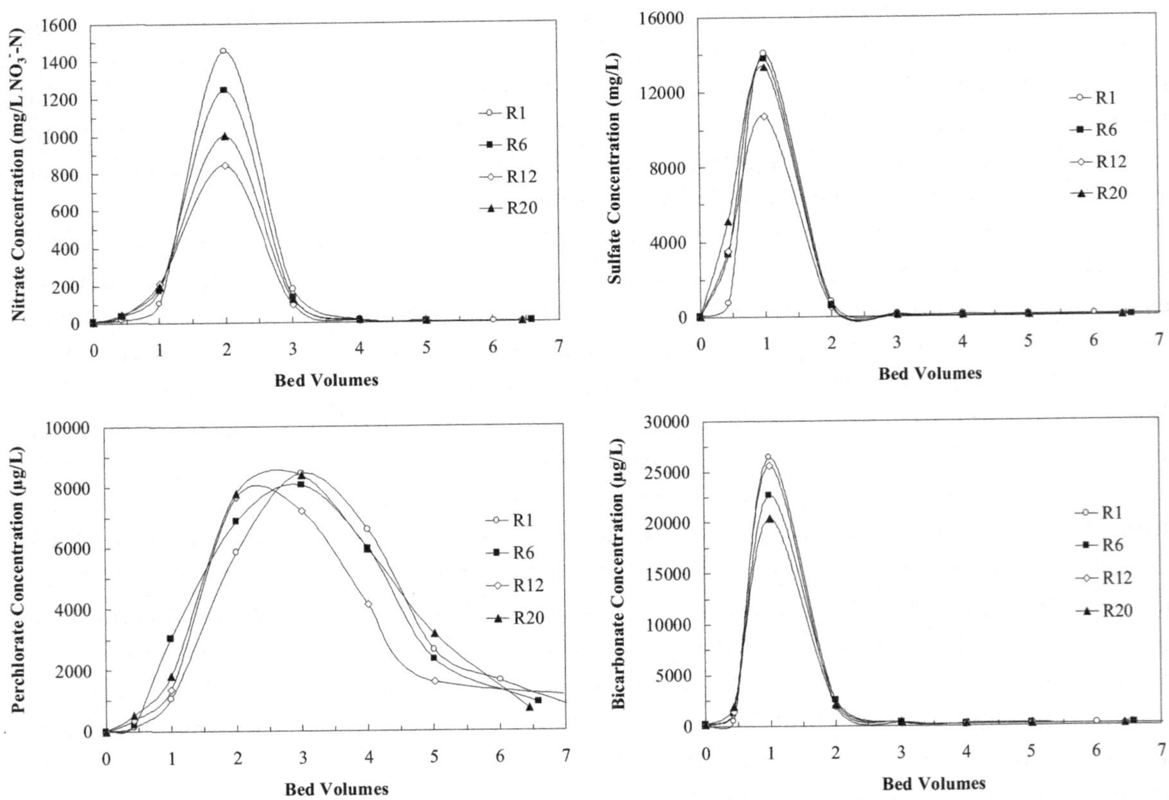

Figure 3.9 Detailed regeneration profiles during baseline testing (polyacrylic resin)

Table 3.2
Summary of perchlorate recovery during baseline evaluation (polyacrylic resin)

Run #	Exhaustion (mg ClO_4^- loaded)	Regeneration (mg ClO_4^- displaced)	ClO_4^- recovery (%)
R1	262	262	100
R6	254	259	102
R12	268	245	91
R20	291	280	96

required. This complete regeneration, however, would only be approximately twice the normal regeneration salt loading rate based on the initial resin evaluation.

From the initial evaluation of the polyacrylic resin, the perchlorate capacity of the resin was found to be approximately 500 mg given this particular water quality. However, analysis of the perchlorate data obtained during selected exhaustion and regeneration profiles (summarized in Table 3.2) revealed that an average of 269 mg of ClO_4^- was adsorbed on the resin. The average perchlorate removed during these regenerations was 97 percent. Although this confirmed that a salt loading rate of 25 lb/ft^3 would be sufficient to minimize the accumulation of perchlorate, it was estimated that a more robust full regeneration would be required after approximately 30 cycles.

Polystyrene 6 Percent Sodium Chloride Performance

Using the operating conditions determined during the initial resin evaluation, the polystyrene resin was exhausted for 625 BV at a flowrate of 1.7 gpm, and counter-current regenerated with 35 lbs/ft^3 using 6 percent NaCl brine. Twenty exhaustion and regeneration cycles were performed to establish baseline performance. Figure 3.10 displays the exhaustion profiles of perchlorate, nitrate, sulfate, chloride, and bicarbonate observed during selected runs (Run 2, 7, 13, and 20). Effluent analysis for the other runs indicated similar performance. As seen in these anion profiles, a high degree of repeatability was observed throughout the baseline testing. This consistency indicated that bicarbonate and nitrate were not accumulating on the resin.

Counter-current regeneration was carried out at a flow rate of 0.37 gpm for 9 BV (salt loading rate of 35 lbs/ft^3) using a 6 percent NaCl brine. Figure 3.11 shows a comparison of selected anions (perchlorate, nitrate, sulfate, and chloride) observed during selected regenerations (Runs 2, 7, 14, and 20). Nitrate and sulfate were displaced from the resin to below their detection limits after 7 BV. At the end of the regeneration cycles, an average perchlorate concentration of 2400 µg/L was still eluting from the resin in the regenerant brine. For all the cycles, the chloride concentration increased for the first 3 BV and then remained steady at 3700 mg/L for the remainder of the regeneration.

A closer look at the area under the regeneration curves for each individual anion (Figure 3.12) revealed that the total mass and elution time of anions adsorbed by the resin during selected runs were consistent. The lack of a consistent increase of peak height indicated that without brine reuse, the anions of interest were not accumulating on the resin within the twenty baseline cycles. The consistent shape and elution times of the measured anions also indicated that the resin bed was not being disrupted during the counter-current regeneration, and that the anions not

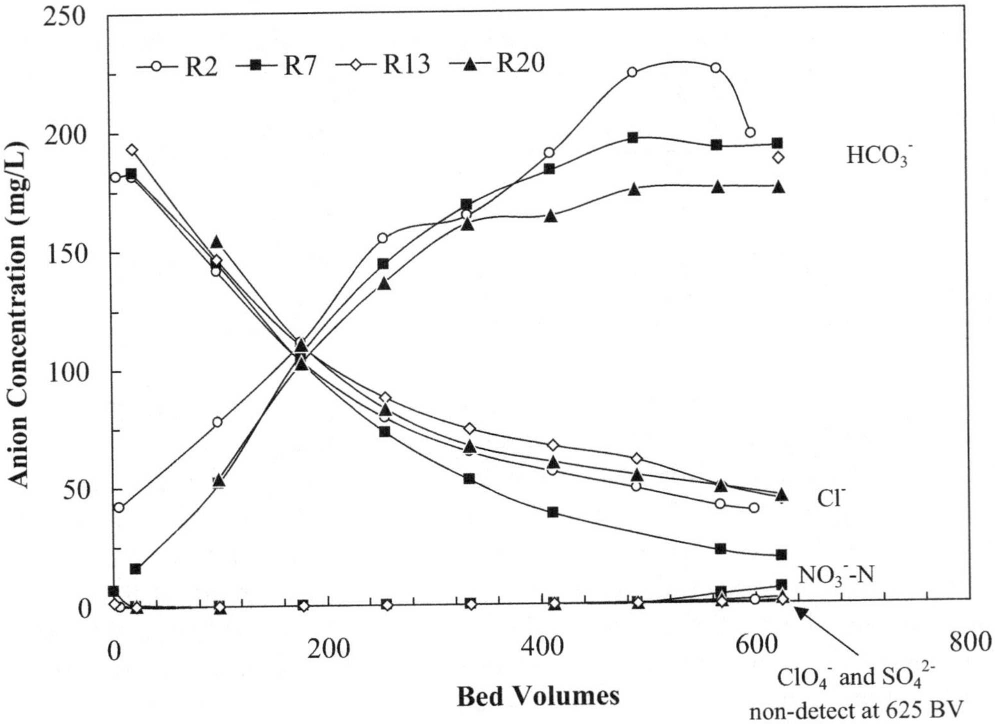

Figure 3.10 Comparison of exhaustion profiles during baseline testing (polystyrene resin)

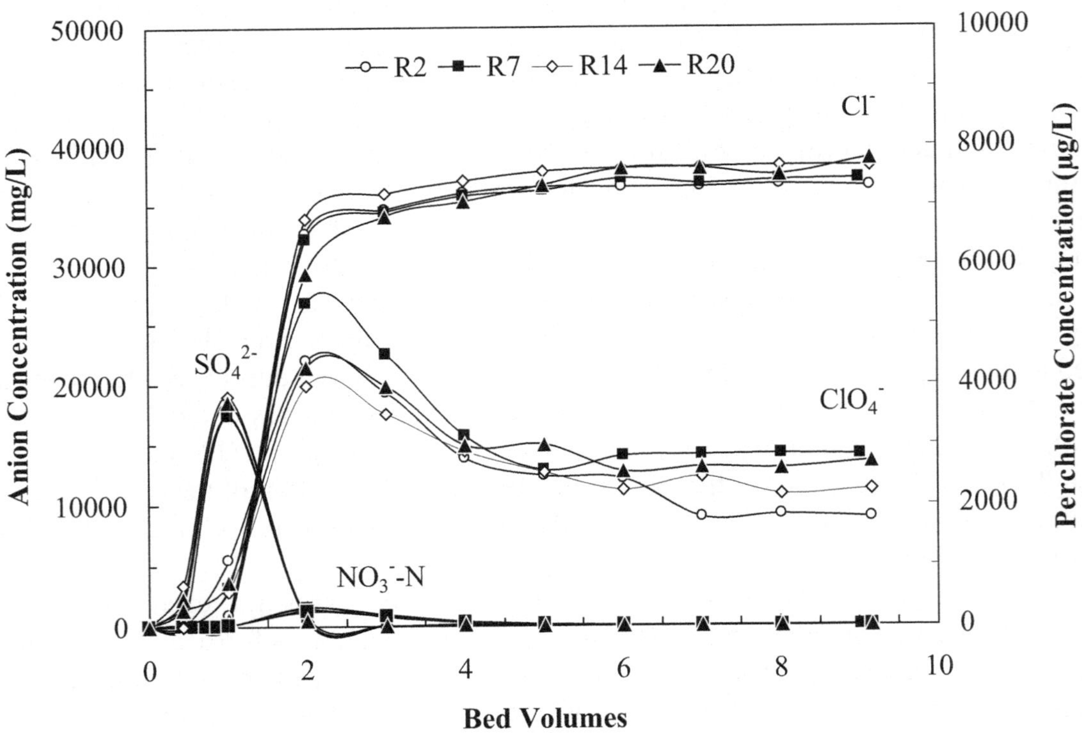

Figure 3.11 Comparison of regeneration profiles during baseline testing (polystyrene resin)

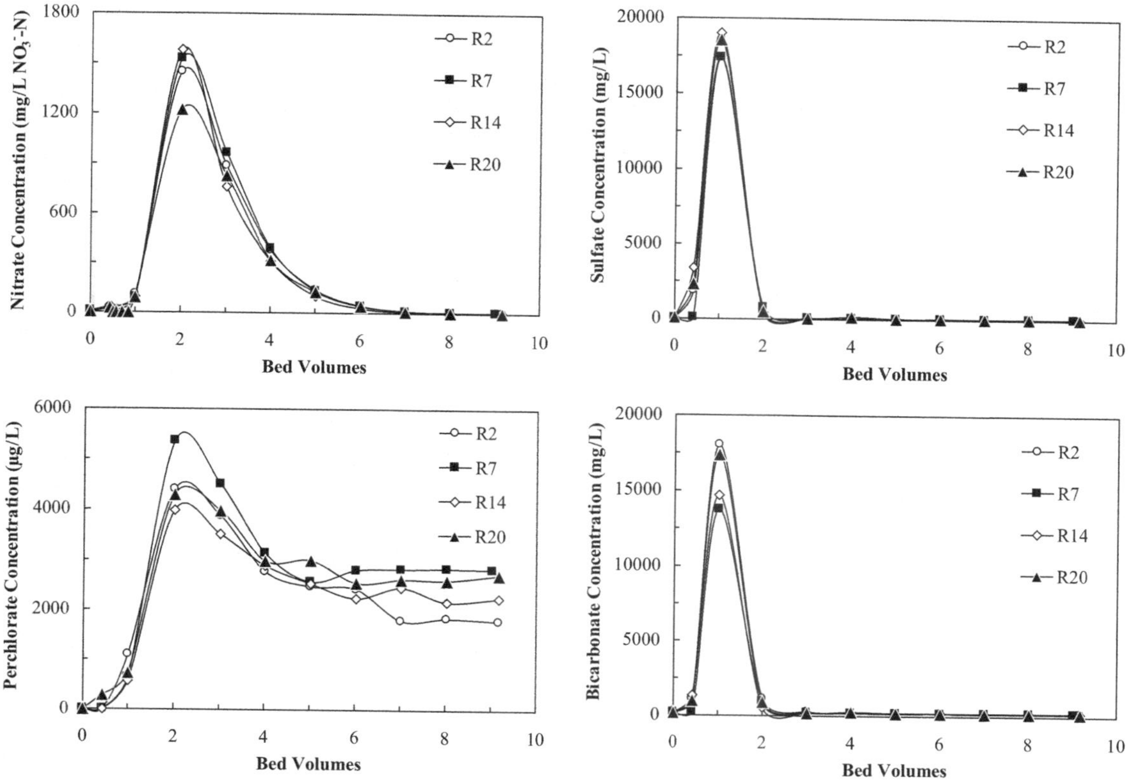

Figure 3.12 Detailed regeneration profiles during baseline testing (polystyrene resin)

routinely measured were not accumulating to level that would impact the ability of the resin to remove perchlorate or nitrate, within the 20 baseline cycles.

From the initial evaluation of the polystyrene resin, the perchlorate capacity of the resin was found to be approximately 3400 mg. However, analysis of the perchlorate data obtained during selected exhaustion and regeneration profiles (summarized in Table 3.3) revealed that an average of 369 mg of ClO_4^- was adsorbed on the resin and up to 72 percent of the perchlorate was removed when a regeneration was performed. This confirmed that a salt loading rate of 35 lb/ft^3 would be sufficient to minimize the accumulation of perchlorate for approximately 30 cycles, at which point a more robust regeneration would be required.

Polyacrylic 3 Percent Sodium Chloride Performance

Previously, baseline performance was established for the polystyrene resin using a 6 percent NaCl regeneration. As the initial biological studies indicated that a culture capable of quickly reducing both perchlorate and nitrate in a 6 percent brine resisted acclimation (see Chapter 5), additional baseline performance was established using a 3 percent NaCl regeneration. Using the same resin exhaustion conditions as used during the 6 percent baseline testing, the polyacrylic resin was exhausted for 400 BV at a flowrate of 1.7 gpm.

Ten exhaustion and regeneration cycles were performed to verify 3 percent NaCl baseline performance. The following Figure 3.13 summarizes the exhaustion profiles of perchlorate,

Table 3.3
Summary of perchlorate recovery during baseline evaluation (polystyrene resin)

Run #	Exhaustion (mg ClO_4^- loaded)	Regeneration (mg ClO_4^- displaced)	ClO_4^- recovery (%)
R2	347	211	61
R7	395	258	65
R14	396	214	54
R20	337	243	72

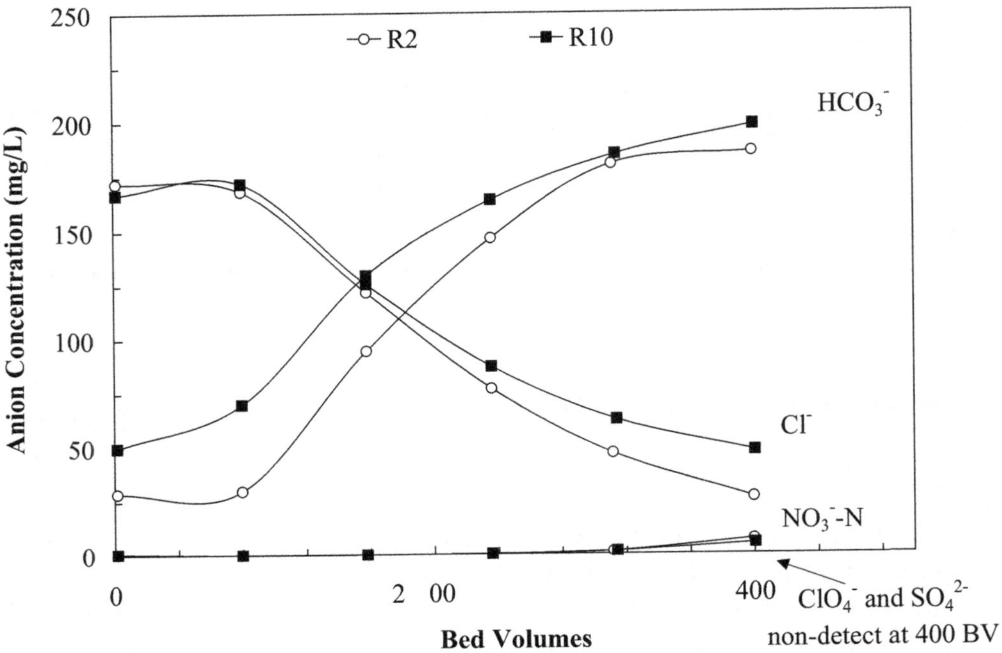

Figure 3.13 Comparison of exhaustion profiles during 3 percent brine baseline testing (polyacrylic resin)

nitrate, sulfate, chloride, and bicarbonate observed during selected runs (Run 2 and 10). Effluent analysis for the other runs indicated similar performance. As seen in these anion profiles, perchlorate and sulfate were consistently below detection limits in the final effluent sample at 400 BV. In addition, nitrate was maintained below the 10 mg-N/L MCL with an average final effluent concentration of 6.5 mg-N/L.

Counter-current regeneration was carried out at a flow rate of 0.37 gpm for 13.5 BV (salt loading rate of 25 lbs/ft^3) using a 3 percent NaCl brine. Figure 3.14 shows a comparison of selected anions (perchlorate, nitrate, sulfate, and chloride) observed during selected regenerations (Runs 2 and 10). Nitrate and sulfate were displaced to below their detection limits from the resin by 7 BV. At the end of the regeneration cycle, perchlorate levels in the final brine samples were below analytical detection limits for all 10 cycles.

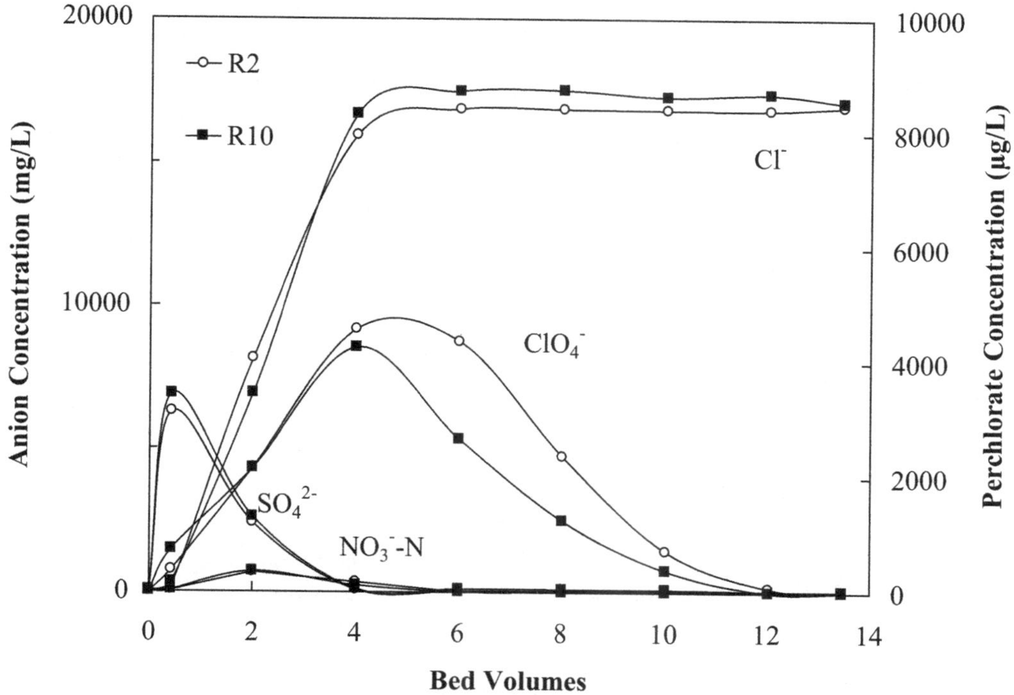

Figure 3.14 Comparison of regeneration profiles during 3 percent brine baseline testing (polyacrylic resin)

Interestingly, for the same salt loading rate, the perchlorate concentration in brine samples taken at 13.5 BV were below detection limits with a 3 percent brine but was approximately 700 µg/L during the 6 percent regeneration. This improved efficiency indicated that regeneration was not instantaneous. At full-scale, a more efficient process might include employing a 6 percent brine with a brief soak period to help address the exchange kinetics.

Heated Regeneration

The effect of heating on the efficiency of regeneration was evaluated for the polystyrene resin. Based on Phase I bench testing results, heating of the regenerant was shown to reduce the amount of salt required for regeneration of polystyrene resins and consequentially reduce episodes of leaching especially when using partial regeneration. Following the baseline testing protocols, fresh polystyrene resin was operated and regenerated at three different temperatures (20, 40, and 60°C) using 9 BV of 7.75* percent NaCl (45 lbs/ft^3 salt loading rate). The temperature of the regenerant solution was increased to the desired temperature using a heat exchanger. During regeneration, a negligible amount heat loss was detected across the resin bed. The differential between the temperature of the influent regenerant brine and the spent brine was less than 2°C.

* The targeted brine composition was 6 percent NaCl, based on previous salt amendments used to prepare brine solutions in the baseline testing. However, analysis of the prepared brine used for heated regeneration experiments revealed that the NaCl concentration was higher than anticipated at 7.75 percent.

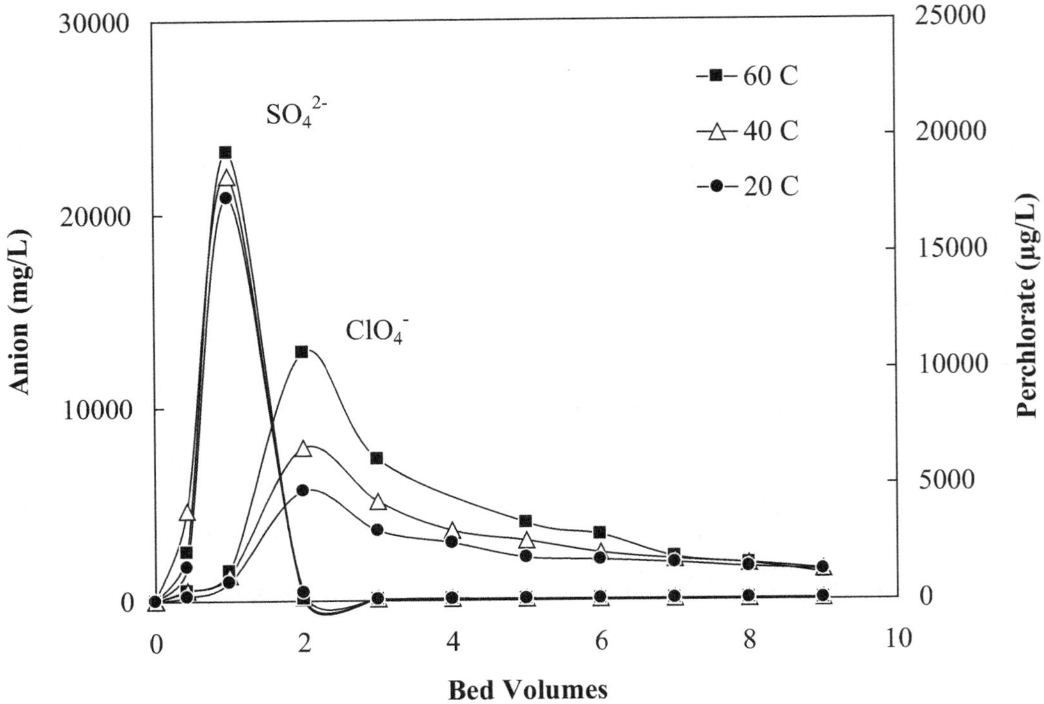

Figure 3.15 Heated regeneration profiles of perchlorate and sulfate (polystyrene resin)

Figure 3.15 and Figure 3.16 show the heated regeneration profiles of sulfate, perchlorate, bicarbonate, and nitrate. A dramatic improvement in the efficiency of heated regeneration was observed at 60°C as compared to 20°C and 40°C. Retention times of all chromatographic peaks remained consistent between the three regeneration temperatures. A closer look at perchlorate revealed that the peak height more than doubled between the ambient 20°C and 60°C conditions. Since the area under the chromatographic peak represents the effective mass of anion removed from the resin during regeneration, it can be seen that the mass of perchlorate removed increased with temperature using the same salt loading rate. To verify this observation, the concentrations of all anions found in the composite brine are presented in Figure 3.17. Perchlorate measured in the composite brine after the 60°C regeneration was nearly 3500 μg/L compared to the 2000 μg/L using 20°C. This improvement can be attributed to the altered equilibrium and improved kinetics at the elevated temperatures.

Although nitrate and sulfate were removed to below detection limits before the completion of regeneration at 20°C, the equilibrium at the elevated temperatures allowed for additional displacement of these anions from the resins. Summarized in Figure 3.18, displacement of nitrate and sulfate to below detection limits during 20°C regeneration occurred at 8 and 7 BV, respectively. During a heated regeneration, however, displacement of both nitrate and sulfate occurred much earlier. For the 60°C regeneration, nitrate and sulfate were displaced from the resin after 6 and 4 BV of regeneration, respectively. In addition, the effective mass of both nitrate and sulfate removed increased as the regeneration temperature was increased, as seen in increasing peak areas of Figure 3.15 and Figure 3.16.

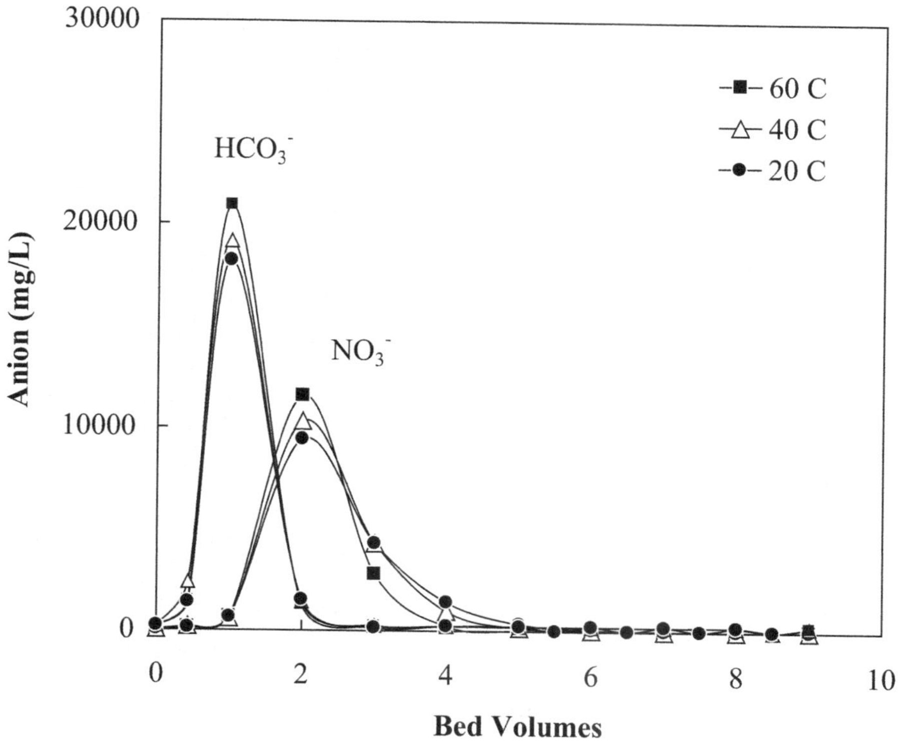

Figure 3.16 Heated regeneration profiles of bicarbonate and nitrate (polystyrene resin)

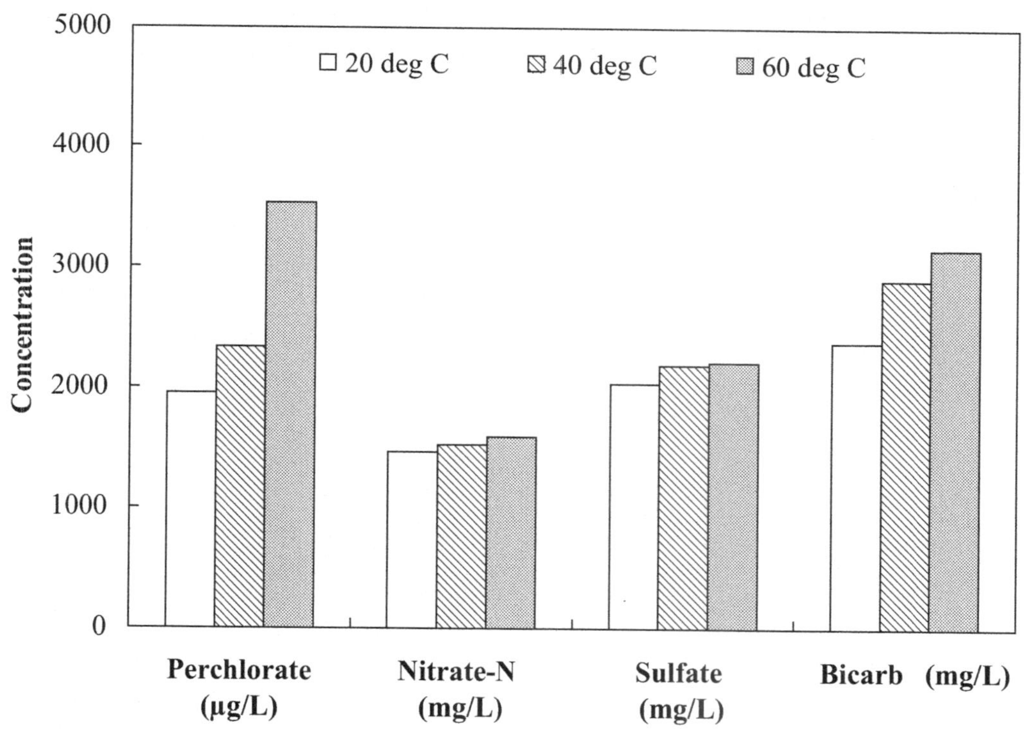

Figure 3.17 Composite brine anions for heated brine regeneration (polystyrene resin)

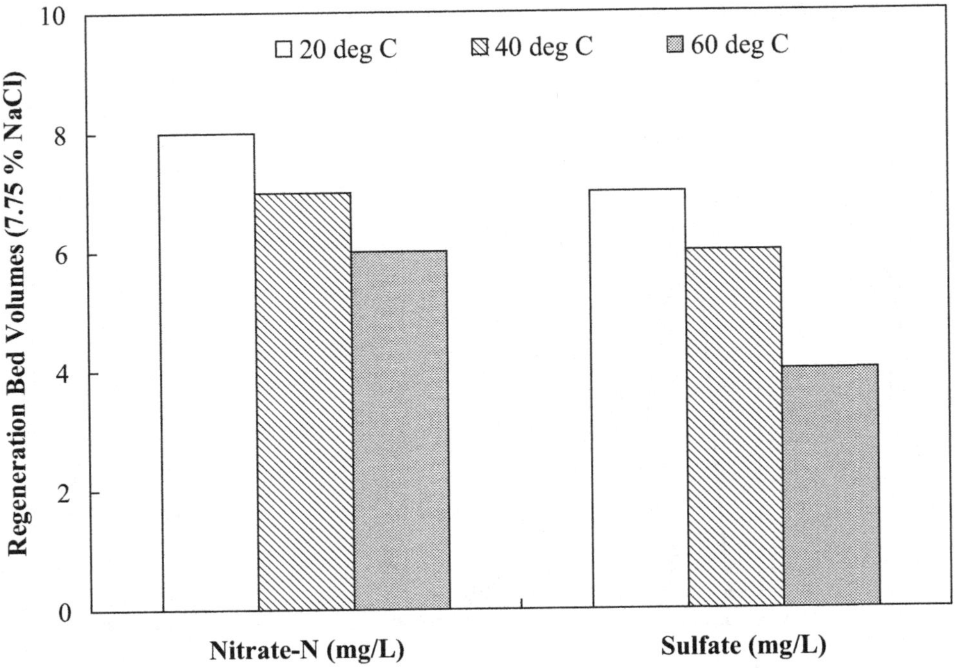

Figure 3.18 Comparison of bed volumes required for complete displacement of nitrate and sulfate during heated regeneration

Apparent from these data, significant improvements in salt usage efficiency were observed at elevated temperatures. Economic analyses would be necessary to determine under what conditions heated regeneration should be considered to capitalize on the increases in regeneration efficiency.

CHAPTER 4
BRINE TREATMENT AND REUSE

The success of the perchlorate ion-exchange process with reuse of treated spent brine depends on the perchlorate level in the recycle brine following treatment. Perchlorate is a monovalent ion, which is not subject to selectivity reversal in the high ionic strength of the NaCl brine used for regeneration. Owing to its inherent high affinity and lack of selectivity (affinity) reversal, a significant portion of any perchlorate remaining in the spent brine will re-adsorb (exchange) onto the resin during regeneration. Then, on the subsequent exhaustion cycle, some of the re-adsorbed perchlorate will exchange for chloride and leak into the column effluent immediately at the start of the run. Consequently, it is in critical to reduce the perchlorate level in the brine to avoid subsequent leakage above the 4-μg/L CaDHS Advisory Action Level.

Table 4.1 presents an analysis utilizing the University of Houston EMCT-Windows model runs with varying amounts of perchlorate in a 6 percent brine for the regeneration of the polyacrylic resin. Based on the trend, the maximum allowable concentration of perchlorate in the treated 6 percent brine was approximately 0.8 mg/L (0.4 mg/L for a 3 percent brine). Manipulation of the model indicated that the equivalent fraction of perchlorate in the early leakage following regeneration would be the same as the equivalent fraction of perchlorate in the treated brine. Therefore, the perchlorate leakage can be quickly estimated as the product of the equivalent fraction perchlorate in the brine and the total concentration of ions in the feed water.

At pilot-scale, brine treatment and reuse was evaluated according to the flow schematic shown in Figure 4.1. Spent brine generated from the pilot ion exchange column was stored in the spent brine tank. Perchlorate and nitrate concentrations were typically 3000 μg/L and 270 mg-N/L, respectively, in 6 percent brine (1000 μg/L and 130 mg-N/L for 3 percent). Once the batch of brine had been collected, it was processed by the brine treatment system (biological, chemical, or electrolytic). Assuming a safety factor of 100 percent, the treatment goal of perchlorate in the 6 percent treated brine was 400 μg/L or less (200 μg/L for 3 percent brine). Treated brine was stored in the treated brine tank and used for the subsequent regeneration process.

Table 4.1
Perchlorate leakage estimates based on UH EMCT-Windows program

ClO_4 in brine, mg/L	Eq fraction, X_{ClO4} in brine after treatment	Eq fraction, Y_{ClO4} on resin after regeneration	ClO_4^- leakage on next run, meq/L	ClO_4^- leakage on next run, μg/L
0.100	10^{-6}	5.52×10^{-6}	0.000005	0.5
1.00	10^{-5}	5.51×10^{-5}	0.00005	5
10	10^{-4}	5.50×10^{-4}	0.00051	51
100	10^{-3}	5.48×10^{-3}	0.0051	510
1000	10^{-2}	5.26×10^{-2}	0.051	5100

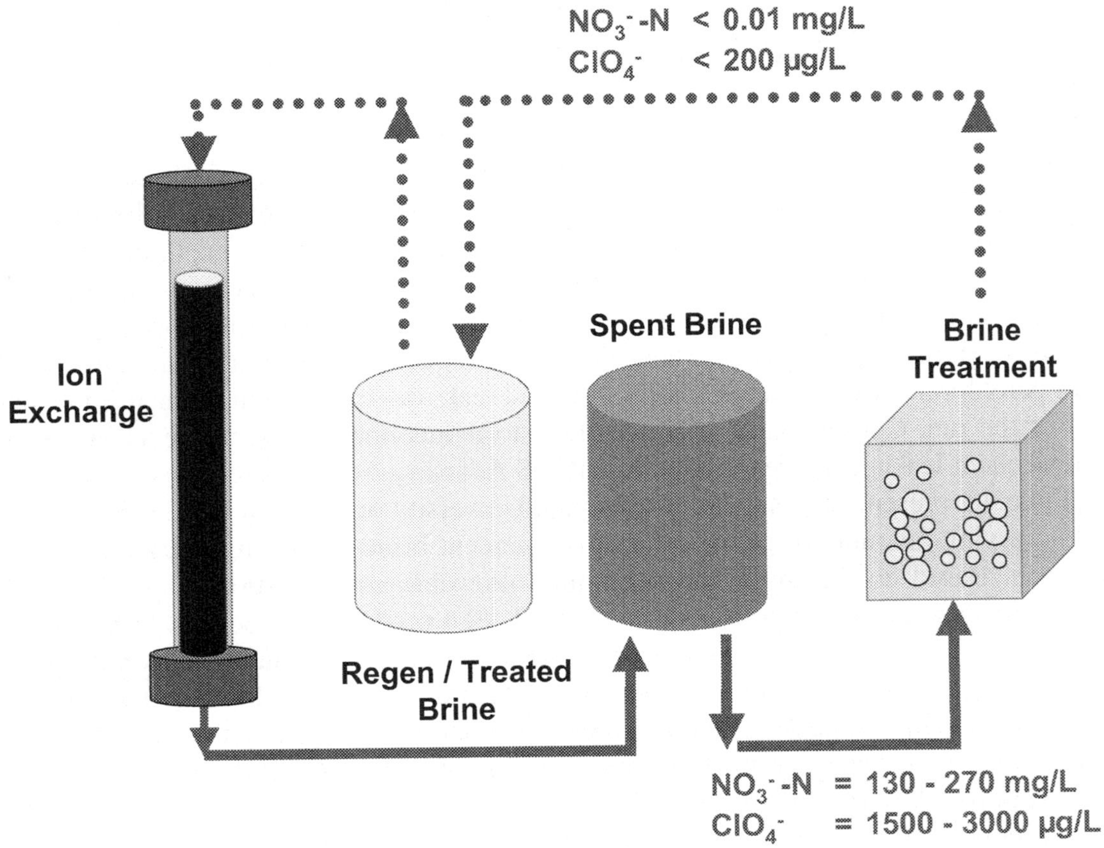

Figure 4.1 Brine treatment and reuse flow schematic

BIOLOGICAL BRINE TREATMENT PROCESS

The biological brine treatment system, shown in Figure 4.2, utilized a marine mud inoculum to biologically reduce both perchlorate and nitrate in a 3 percent NaCl regenerant brine solution. The pilot-scale reactor operated as an anaerobic sequencing batch reactor (SBR) had the following features:

- Hermetically sealed high density polypropylene tank with a self-closing air break to release internal pressure developed during biological reduction and nitrogen sparging
- Coarse diffuser plate located at the bottom of the reactor for oxygen stripping and nitrogen sparging to assist with the suspension of biomass
- Integrated shaft mixer to suspend biomass
- Capability for closed loop recycling of biomass from a port located in the bottom of the reactor with return line to the top
- Dedicated fill port located at top of the reactor for feeding brine and supplying chemicals
- Multiple drain and sample ports located on the sides and bottom of the reactor for operational flexibility as a sequencing batch reactor

Figure 4.2 Biological brine treatment system

Operation of the biological brine treatment system followed the conventional fill-and-draw procedure: (1) fill, (2) mix and react, (3) settle, and (4) draw. During the pilot-scale fill operation, spent brine generated during the resin regeneration process was diverted directly to the SBR at a flowrate of 0.37 gpm, during which time, nitrogen sparging and reactor mixing was used encourage the stripping of oxygen. Once the reactor had completed filling, acetic acid was supplied to the culture, based on the stoichiometric requirement of acetate to perchlorate and nitrate, as an electron donor for the reduction process. While a 200 mg/L excess acetate concentration was maintained in the culture, this passed through the regeneration process and thus, did not have to be added with each cycle.

The SBR contents, now containing the perchlorate- and nitrate-laden brine amended with acetate, were mixed and allowed to react. Once the perchlorate and nitrate were reduced to the treatment goals (NO_3^- < 0.01 mg/L and ClO_4^- < 200 µg/L), the SBR entered the settling phase. The suspended biomass in the treated brine was allowed to settle for 1 to 2 hours. After settling, approximately 65 percent of the reactor supernatant was drawn, filtered, and collected in the regeneration tank. Additional NaCl was added to the treated regenerant brine solution to readjust the chloride concentration before being reused for the following regeneration of the exhausted ion exchange resin. Additional details, performance, and operation and maintenance issues will be discussed further in Chapter 5.

Figure 4.3 Chemical brine treatment system

CHEMICAL BRINE TREATMENT PROCESS

The physical/chemical brine treatment system (PNDM - Perchlorate and Nitrate Destruction Module) was provided by Calgon Carbon Corporation and is shown in Figure 4.3. The PNDM employed a high-pressure and high-temperature catalytic process to reduce the nitrate and perchlorate in the spent brine. The PNDM used a chemical reductant (ammonia) with a stoichiometric dose based on the measured concentrations of nitrate and perchlorate in the spent brine.

The basic operational parameters included:

- Flowrate = 150 mL/min
- EBCT = 10 min
- Spent brine batch volume = 66 L
- Treatment time = 7 hours
- Reactor Temperature = 230°C
- Reactor Pressure = 600 psi
- Excess ammonia residual = 0.5 g/L

Successful operation of the PNDM required continuous operation. As the pilot testing was focused on batches of brine, a virgin 6 percent NaCl brine with 0.5 g/L ammonia was recirculated through the system. Once the spent brine from the pilot ion exchange system had been collected, the nitrate and perchlorate concentrations were measured. A stoichiometric dose of ammonium

Figure 4.4 Electrolytic brine treatment cell

chloride was added to the spent brine (note: the spent brine already contained approximately 0.5 g/L excess ammonia applied during the first cycle). Once dosed with ammonia, the feed was switched to the spent brine from a recirculating virgin brine. The spent brine was then pressurized to 600 psi and fed to the reactor on a continuous basis (150 ml/min). The reactor heated the brine to 230°C before it reached the fixed catalyst bed. Nitrate and perchlorate were reduced as they passed through the reactor. The treated brine and nitrogen gas generated during the reduction of nitrate exited the reactor to be cooled through a heat exchanger before being discharged to the treated brine tank, ready to be used for the next regeneration without any further treatment. Additional details, performance, and operation and maintenance issues will be discussed further in Chapter 6.

ELECTROLYTIC BRINE TREATMENT PROCESS

The electrolytic brine treatment system (IXL™) was provided by Ionex Ltd. (see Figure 4.4). This process utilized a simple bipolar electrochemical cell to reduce perchlorate and nitrate to chloride, nitrogen gas and oxygen. Another by-product of this reaction was the generation of hydroxide. Following the electrolytic cell, the only necessary changes in the brine to make it suitable for reuse included pH adjustment and chloride addition. Conceptually, once perchlorate and nitrate had been reduced the brine could be immediately reused. This process did not require the addition of an electron donor as with the biological process, or a reductant as with the physical/chemical treatment process. Additional details, performance, and operation and maintenance issues will be discussed further in Chapter 7.

BRINE TREATMENT AND REUSE TESTING OBJECTIVES

The following three chapters discuss the project team's evaluation of each brine treatment technology at bench- and/or pilot-scale. Detailed experiments were performed with each brine treatment system (biological, chemical, electrolytic) to address the following brine treatment and reuse objectives:

- Evaluate the ability of each brine treatment process to remove perchlorate and nitrate
- Determine the long term operational reliability
- Identify operation and maintenance requirements
- Determine how reusing treated brine may affect resin performance
- Identify any by-products that may be accumulated during reuse

CHAPTER 5
BIOLOGICAL BRINE TREATMENT

Initial studies during Phase I investigations revealed that selected cultures exposed to both perchlorate and nitrate, preferentially reduced the nitrate and only reduced perchlorate after long lag periods. As a result, the first two candidate cultures were established to sequentially degrade nitrate and perchlorate in two separate reactors to avoid nitrate inhibition of perchlorate degradation. These organisms, however, were not able to acclimate to 20 g/L of NaCl, rendering them ineffective for the treatment of typical ion exchange brines (at least 30 g/L NaCl).

The use of a marine mud inoculum proved to be more successful in obtaining a useable culture. Unlike the initial tests, the results using this inoculum showed that perchlorate degradation proceeded in the presence of nitrate. Two cultures were then established. One culture was capable of reducing 500 mg/L nitrate and 100 mg/L perchlorate within a 24-hour period in the presence of 30 g/L salt. The second culture (100-mL) was able to reduce 100 mg/L perchlorate in the presence of 60 g/L NaCl within 29 hours. As reported in the Phase 1 final report, these cultures were stable and reduced perchlorate and nitrate simultaneously in a synthetic medium.

This chapter discusses the subsequent Phase II development, utilization, and characterization of the biological brine treatment cultures. The following topics will be discussed:

- Conversion from synthetic medium feed to brine feed
- Culture nutrient requirements
- Impacts on culture performance
- Pilot-scale culture development
- Pilot brine treatment and recycling
- Microbiological analysis of cultures
- Pure culture characterization

CONVERSION FROM SYNTHETIC MEDIUM FEED TO BRINE FEED

The 6 percent NaCl culture described above was initially increased in volume from 100 mL to 1.5 L by transfers to a large bottle containing 500 mL 6 percent NaCl synthetic medium and the addition of 500 mL fresh medium each time the perchlorate was degraded. The composition of this medium is summarized in Table 5.1. Two 1.5-L perchlorate-reducing cultures (one at 6 percent and one at 3 percent) were obtained. These were consistently fed two years with a synthetic medium designed from seawater without sulfate (Table 5.1). In order to ensure that active cultures were maintained, both of these cultures were split, producing four 1.5-L cultures; two adapted to 3 percent NaCl and two adapted to 6 percent NaCl. Next, one 3 percent and one 6 percent culture was switched to a brine feed generated from the pilot-scale ion exchange system described in Chapter 2. The major brine constituents are listed in Table 5.2.

Response of 3 Percent NaCl Culture to Brine Feed

Once the two 1.5-L 3 percent NaCl synthetic medium cultures demonstrated similar performance, one was switched to a spent process brine feed instead of the synthetic seawater medium. For the first four months of maintenance, the culture was incubated at 30°C. Following

Table 5.1
Constituents of synthetic medium

Component	Concentration
$MgCl_2{\cdot}6H_2O$	11 g/L
$CaCl_2{\cdot}2H_2O$	1.4 g/L
KCl	0.72 g/L
NH_4ClO_4	117 mg/L
Sodium Acetate$\cdot3H_2O$	2 g/L
NaCl	60 g/L
Resazurin	0.3 mL/L
$NaHCO_3$*	0.2 g/L
Phosphate Solution*	1.0 mL/L
[50 g/L KH_2PO_4]	
67mM $Na_2S{\cdot}9H_2O$*	5.0 mL/L
Mineral Solution #2†	1.0 mL/L
$(NH_4)_6Mo_7O_{24}{\cdot}4H_2O$	10
$ZnCl_2$	0.05
H_3BO_3	0.3
$FeCl_2{\cdot}4H_2O$	1.5
$CoCl_2{\cdot}6H_2O$	10
$MnCl_2{\cdot}6H_2O$	0.03
$NiCl_2{\cdot}6H_2O$	0.03

* Items added after initial boiling and sterilization.
† Mineral Solution #2 Composition in Deionized Water (g/L).

Table 5.2
Constituents of ion-exchange spent brine (process brine)

Constituent	Concentration
ClO_4^-	3,870 (μg/L)
Cl^-	25,400 (mg/L)
NO_3^-	1,050 (mg/L)
SO_4^{-2}	1,640 (mg/L)

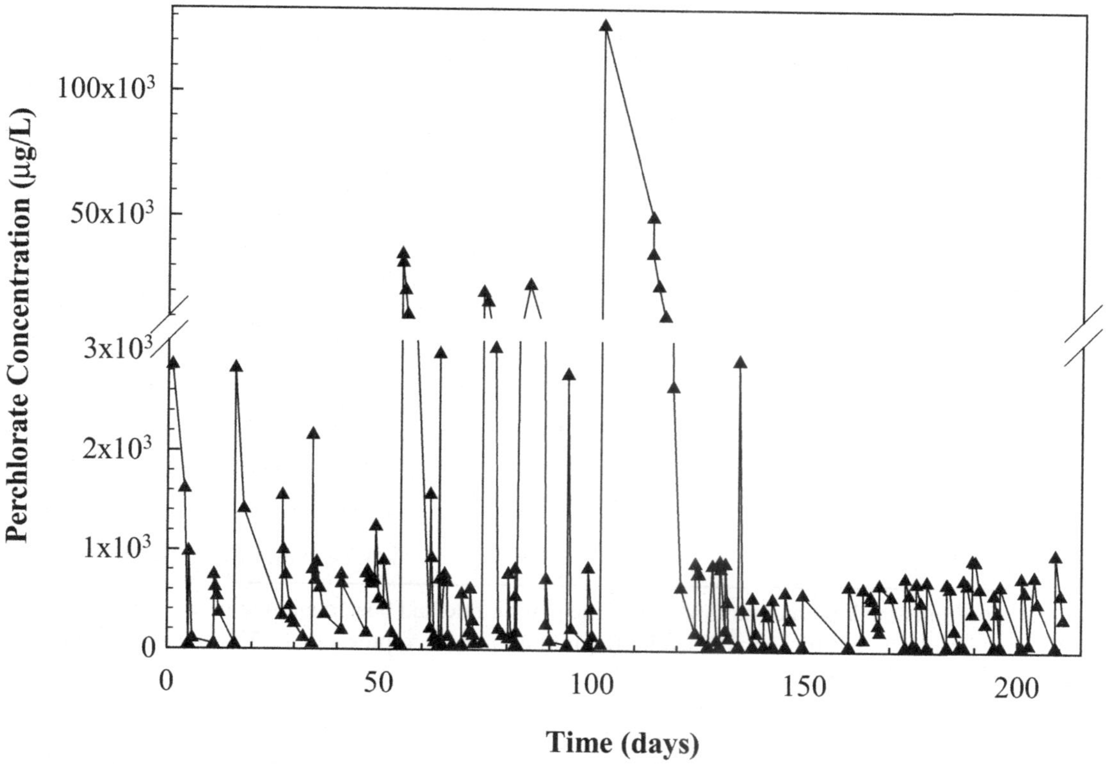

Figure 5.1 Perchlorate reduction in the 3 percent NaCl brine culture

this period, the culture was then maintained at room temperature. The culture was fed as a sequencing batch reactor (SBR) as follows: after the reaction period, the culture was allowed to settle for 1-2 hours, and 40-60 percent of the total volume was replaced with fresh 3 percent NaCl process brine. Acetate (~1 g/L) and resazurin were added with each transfer. Nitrogen gas production from nitrate as well as the perchlorate concentrations were monitored. Figure 5.1 presents the perchlorate concentration in the culture throughout the entire incubation period (8 months).

The feeds or replacements were done once a week during the first 7 cycles. The perchlorate concentration in the fourth through sixth feeds was increased to approximately 3000 µg/L by spiking with additional perchlorate in an attempt to increase perchlorate specific biomass. Visual observations indicated that biomass in the brine culture was not increasing to match the biomass in the synthetic medium culture. Volatile suspended solids (VSS) measurements after the eighth transfer revealed that the culture fed brine had 355 mg/L VSS while the culture fed synthetic medium supported over five times the biomass with a VSS of approximately 2000 mg/L.

During the ninth transfer, the biomass was artificially increased to approximately 930 mg/L VSS by adding 140 mL concentrated biosolids from the synthetic medium culture (Table 5.3). Improvement of perchlorate reduction was noticed with perchlorate being degraded to below 200 µg/L from ~900 µg/L within 47 hours. The perchlorate concentration in the tenth feed was increased to approximately 35 mg/L in an attempt to stimulate perchlorate specific biomass. The results from the eleventh to twenty-second feeds showed that adding more biomass to the culture and spiking with higher concentrations of perchlorate increased the perchlorate degradation

Table 5.3
Volatile suspended solids (3 percent NaCl brine culture)

Feed cycle*	VSS (mg/L)	Change
After 8th cycle	355	
After increase on 9th replacement	930	+575
After 10th cycle	1000	+70 (from after 9th)
After 11th cycle	800	-200 (from after 10th)
After 12th cycle	710	-90 (from after 11th)
Beginning of 13th cycle	540	-170 (from after 12th)
After 13th cycle	637	-73 (from after 12th) +97 (from beginning of 13th)
Beginning of 14th cycle	680	-250 (from 9th replacement)
Beginning of 23th cycle	363	-567 (from 9th replacement)

*After a cycle represents the VSS measurement after the perchlorate had been degraded in that replacement cycle, but before settling and replacement for the next cycle. Beginning of cycle represents the VSS measurement immediately after the replacement procedure but before perchlorate degradation has occurred.

rate. The perchlorate was generally reduced from approximately 1000 μg/L to below 200 μg/L within 24 hours during this time.

Nitrate degradation, monitored by nitrogen (N_2) gas production, was consistent throughout the incubation period. Nitrogen gas production was usually complete within the first twelve hours after each feed cycle. The sulfate present in the brine was not found to inhibit perchlorate reduction nor was sulfate reduced by the culture. Additionally, neither phosphate nor sulfide addition appeared to increase biomass or accelerate perchlorate reduction.

Although the culture was functional, Table 5.3 shows that the biomass was not maintained during transfers. The VSS concentration decreased from 930 mg/L to 363 mg/L between the ninth and twenty-third replacement, respectively. It was hypothesized that a nutrient limitation existed in the brine culture. Experiments to test this hypothesis were conducted during the remainder of the experimental period.

After the twenty-third cycle, a significant portion of the biomass was removed for nutrient addition tests. The remaining biomass was insufficient to maintain stable performance and slowed perchlorate degradation to a 3 to 4 day period. The addition of potassium alone did not improve the rate of perchlorate degradation. The simultaneous addition of magnesium, calcium, and potassium, however, was able to accelerate perchlorate reduction. Unfortunately, this addition only enabled the culture to reduce perchlorate in 2 days, as compared to the less than 24 hours experienced with the culture prior to the biomass removal.

Response of 6 Percent NaCl Culture to Brine Feed

The 6 percent NaCl culture in a synthetic medium was spiked with perchlorate whenever the perchlorate concentration in the culture was measured below detection limits (approximately every third day). With every fifth perchlorate spike, 1 g/L of acetate was added to the culture. During this initial period, the culture was also transferred with every tenth spike of perchlorate

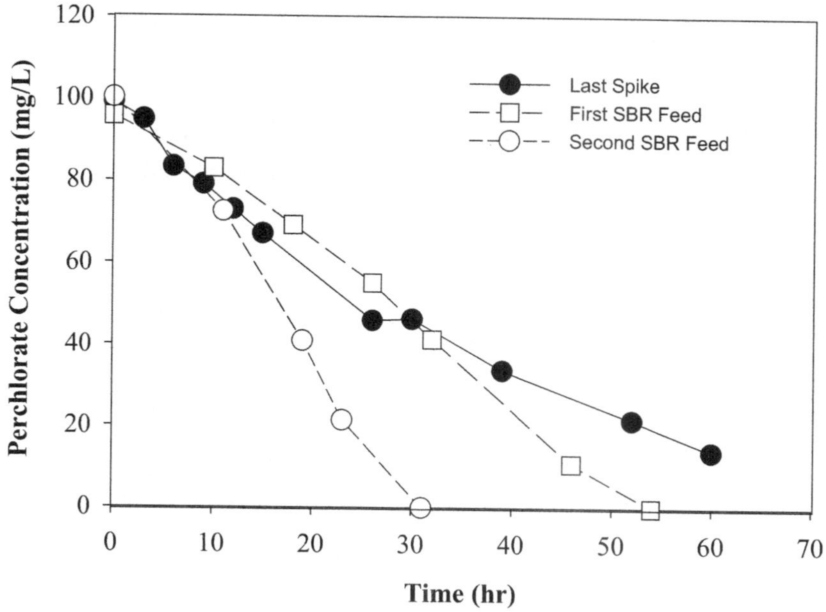

Figure 5.2 Response of 6 percent NaCl culture to sequencing batch reactor conditions

(approximately once per month). At the end of a 4-month period, this 1.5-L culture was subjected to SBR operation using a 40 percent transfer of the medium after settling. Figure 5.2 shows that the culture responded positively to the new procedure demonstrating faster perchlorate removal after each successive SBR cycle.

Once the 6 percent culture was successfully reducing 80 mg/L of perchlorate to below analytical detection limits within 24 hours, the culture was fed spent process ion-exchange brine from the pilot plant. The brine, initially containing 3.9 mg/L perchlorate and 1000 mg/L nitrate, was spiked to 86 mg/L perchlorate and used for a 60 percent transfer. Figure 5.3 shows that perchlorate reduction immediately slowed. The culture was not able to adapt to continued replacements with an unamended 6 percent brine (see Figure 5.4). The perchlorate removal rate slowed with each 60 percent transfer. By the sixth brine replacement, more than 200 hours were required to reduce perchlorate to below detection limits.

Investigation of Brine Effect

One possible reason that the cultures fed actual spent process brine were sustainable was that there were some unknown components present in the process brine that were limiting or inhibiting the growth of the organisms in the culture. Experiments comparing the performance of the cultures when fed with synthetic medium, synthetic brine, and actual spent process brine at 3 and 6 percent NaCl were used to help identify possible reasons for the poor performance of the cultures in the actual spent process brine. The synthetic brine was designed to contain all of the major known components of the actual spent process brine.

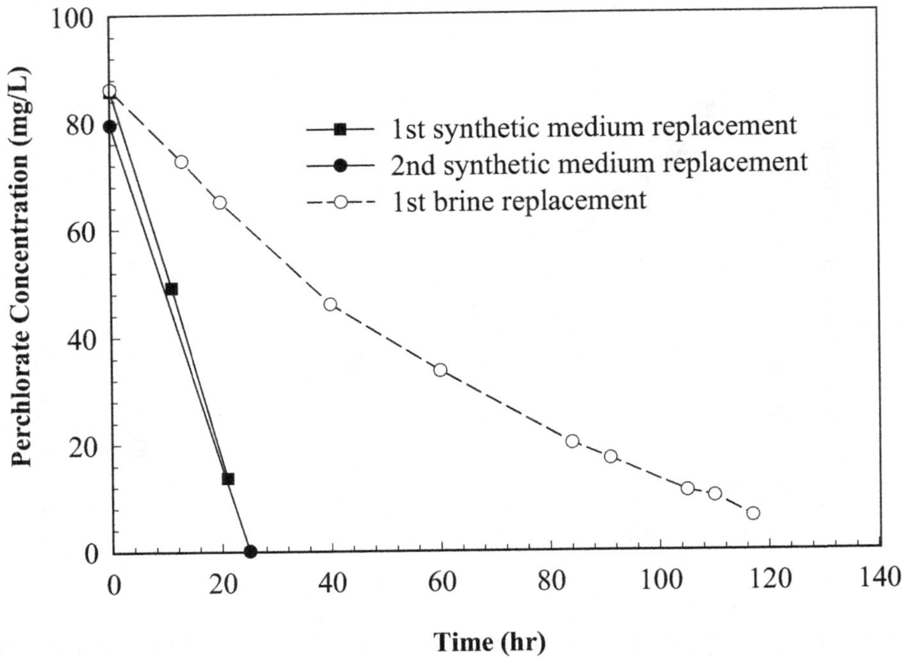

Figure 5.3 Effect of brine addition on perchlorate removal by the 6 percent NaCl culture

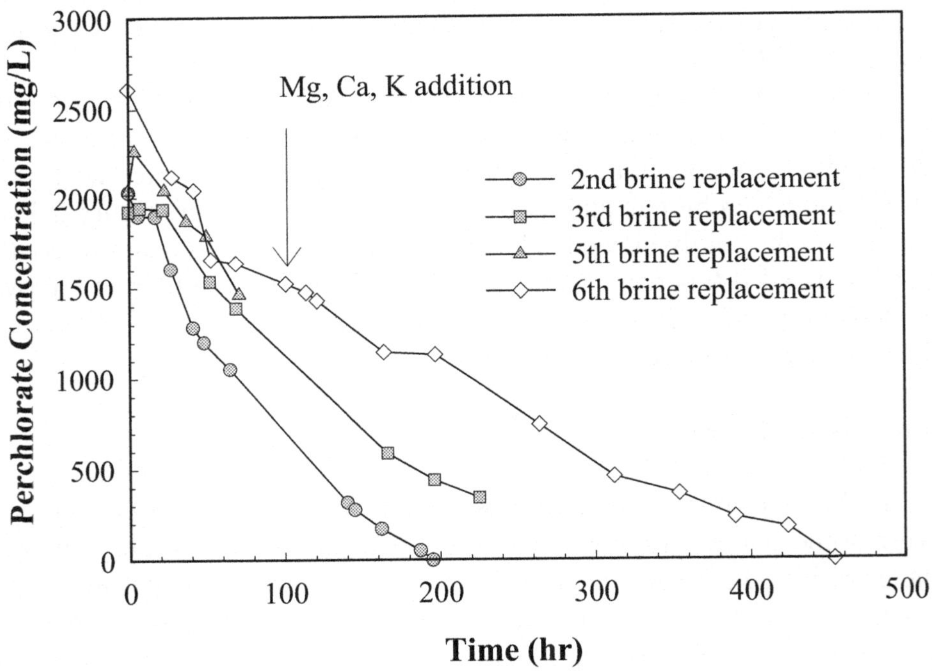

Figure 5.4 Response of 6 percent NaCl culture to additional brine replacement

Table 5.4
Composition of media used in the brine effect experiment (3 percent culture)

Component	3 percent synthetic medium	3 percent synthetic brine	3 percent process brine (20 cycles)
NaCl	30 g/L	30 g/L	30 g/L
		Cations	
Mg^{2+}	1300 mg/L	6.5 mg/L	6.5 mg/L
Ca^{2+}	380 mg/L	24.5 mg/L	24.5 mg/L
K^+	380 mg/L	None	None
NH_4^+	None	33.7 mg/L	2.28 mg/L
		Anions	
NO_3^-	None	511 mg/L	509 mg/L
SO_4^{2-}	None	811 mg/L	800 mg/L
HCO_3^-	145 mg/L	508 mg/L	1960 mg/L
$Na_2S{\cdot}9H_2O$	0.335 mM	None	None
H_2PO4^-	35.6 mg/L	None	None
Trace Metal	Present	None	None

3 Percent Culture Growth Medium

A concentrated 10 percent transfer of the original synthetic medium culture was used to inoculate 100-mL sub-cultures of the three different media in triplicate. The three media used in this experiment were the 3 percent synthetic medium media, 3 percent synthetic brine, and actual spent process brine diluted to 3 percent NaCl (Table 5.4). No obvious differences in perchlorate reduction by the cultures fed the three different media were observed in this first transfer, as shown in Figure 5.5. With a 10 percent transfer, no significant changes were anticipated as most of the ingredients present in the initial inoculum would be present in significant amounts in the inoculum.

After the perchlorate concentration had decreased to below 200 μg/L in this first transfer, a second 10 percent transfer (complete mix) was made with fresh media. Differences in the ability of the sub-cultures to remove perchlorate were noticeable in this second transfer. In the control culture, fed synthetic medium, perchlorate was removed within 10 hours while the cultures fed with either the synthetic or actual brine required approximately 120 hours to degrade the same amount of perchlorate.

After the organisms had reduced the perchlorate to below 200 μg/L in the second transfer, all of the cultures were settled as in a SBR and the supernatant, 70 mL, was again replaced with fresh media of the same composition. Similar performance was observed for both the synthetic medium and synthetic brine after the third transfer as compared to the second transfer. The performance of the sub-cultures fed actual process brine, however, continued to worsen. Instead of requiring 120 hours to degrade perchlorate to below detection, more than 172 hours were required (see Figure 5.6). Unfortunately, these results did not clearly indicate whether or not: 1) something present in the synthetic medium was missing from the synthetic or actual brines; 2) something in the brines was inhibitory to the culture; or 3) a combination of both.

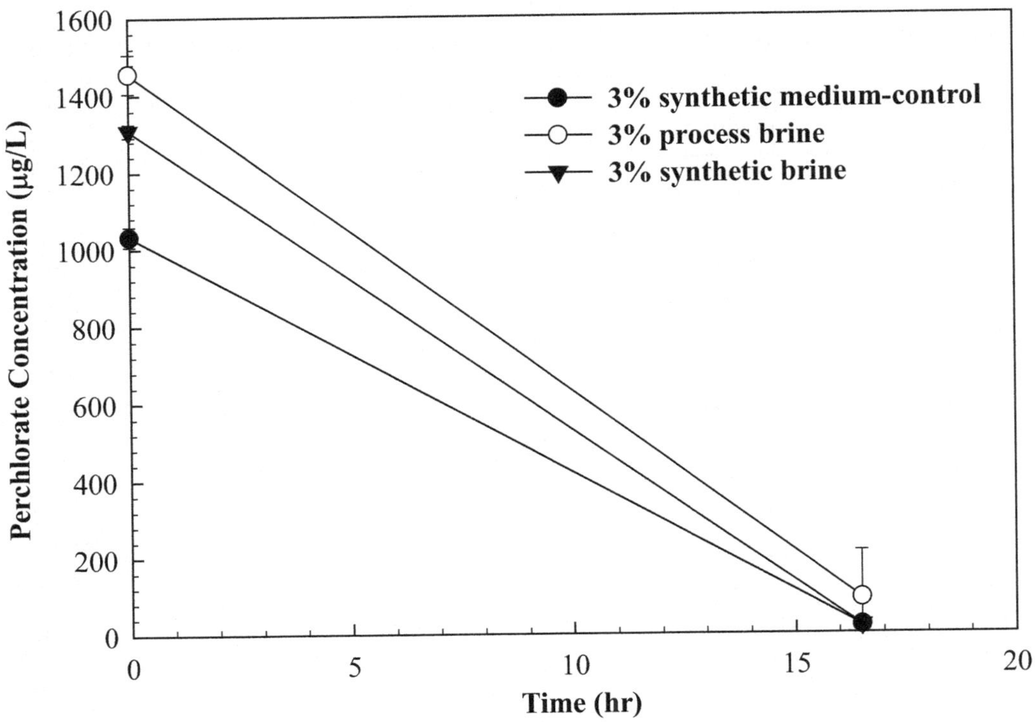

Figure 5.5 Effect of brine on perchlorate reduction: 3 percent culture 1st transfer

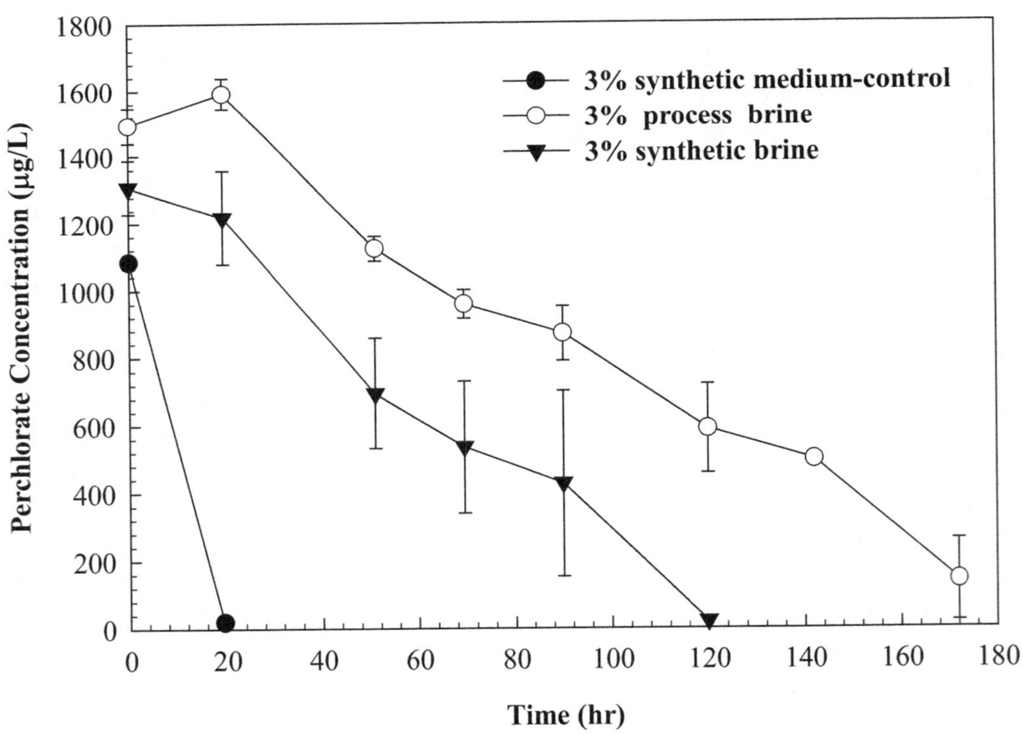

Figure 5.6 Effect of brine on perchlorate degradation: 3 percent culture 3rd transfer

Table 5.5
Composition of media used for the 6 percent brine effect experiment

Components	6 percent synthetic medium	6 percent synthetic brine	6 percent MWH brine
NaCl	60 g/L	60 g/L	60 g/L
	Cations		
Mg^{2+}	1300 mg/L	13 mg/L	13 mg/L
Ca^{2+}	380 mg/L	49 mg/L	49 mg/L
K^+	380 mg/L	None	None
NH_4^+*	None	67 mg/L	4.55 mg/L
	Anions		
NO_3^-	None	1021 mg/L	1018.6 mg/L
SO_4^{2-}	None	1622 mg/L	1600 mg/L
HCO_3^-	145 mg/L	1016 mg/L	3920 mg/L
$Na_2S{\cdot}9H_2O$	0.335 mM	None	None
$H_2PO_4^-$	35.6 mg/L	None	None
Trace Metal	Present	None	None

* NH_4Cl was added in 3 percent synthetic brine in order to make sure that nitrogen is not limiting the growth of the culture.

Based on the previous experience with the parent 3 percent NaCl brine culture, and since the synthetic and process brine subcultures had significantly lower concentrations of magnesium, calcium and potassium, each of these cations were added to the synthetic and process brine sub-cultures, 260, 380 and 380 mg/L respectively. Immediate restoration of the sub-cultures to rapid perchlorate degradation was not observed. Based on observations of the apparent amount of biomass in each of the sub-cultures, it was likely that an insufficient amount of biomass was present to immediately recover performance, regardless of how the cultures were amended.

6 Percent Culture Growth Medium

These experiments with the 3 percent culture were repeated with a set of 6 percent NaCl sub-cultures. The composition of the three different media utilized in these experiments are summarized in Table 5.5. In contrast to the 3 percent NaCl experiments, an immediate difference was observed between the synthetic medium and the synthetic and actual process brine sub-cultures with the first transfer (see Figure 5.7).

With the second transfer, a continued decrease in performance was observed with all of the sub-cultures, presumably due to a loss of biomass during the transfers. The synthetic medium required approximately 50 hours to degrade the perchlorate after this transfer, as compared to less than 16 hours with the first transfer. The synthetic and process brine sub-cultures time required to degrade the perchlorate increased from approximately 70 hours with the first transfer, to between 90 and 120 hours with the second transfer.

The results of the third transfer are presented in Figure 5.8. This time the performance of the synthetic medium sub-cultures was consistent with the previous transfer. The synthetic and

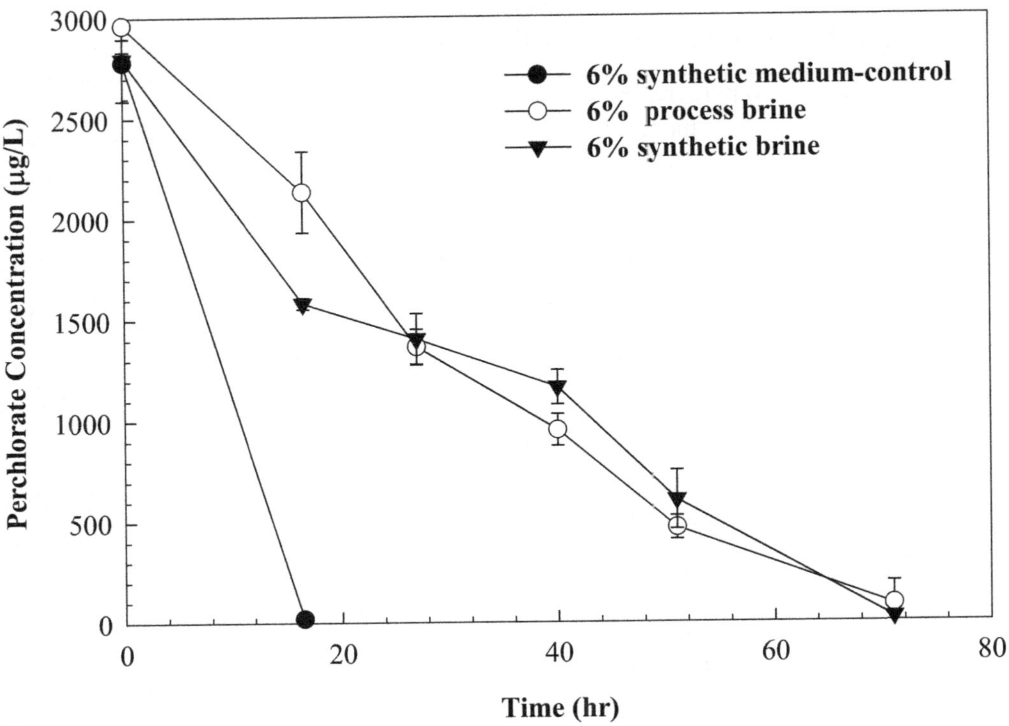

Figure 5.7 Brine effect on perchlorate degradation: 6 percent culture 1st transfer

process brine sub-cultures, however, continued to lose their ability to rapidly degrade perchlorate. Even after 215 hours, only 30 to 50 percent of the perchlorate had been degraded.

As with the 3 percent sub-culture medium experiments, the addition of magnesium (260 mg/L), calcium (380 mg/L) and potassium (380 mg/L) did not improve or stabilize the performance of either the synthetic or process brine sub-cultures.

NUTRIENT REQUIREMENTS

To help determine why the cultures did not perform as well in brine (actual process or synthetic) as compared to the synthetic growth medium initially used to develop the cultures, a series of nutrient requirement experiments were performed. These experiments looked at the impacts of phosphate, magnesium, calcium, potassium and acetate.

Phosphate

Phosphate is a required nutrient for all biological systems; it is an important component of DNA, RNA, and membranes. Consequently, it was thought that the low levels of phosphate in the brine could be a nutrient limitation. To evaluate this, two series of sub-cultures were prepared with the 6 percent synthetic medium shown in Table 5.5, with phosphate being omitted from one of the series of sub-cultures. Unexpectedly, phosphate appeared to be inhibitory to the performance of the culture, slowing perchlorate degradation, as shown in Figure 5.9.

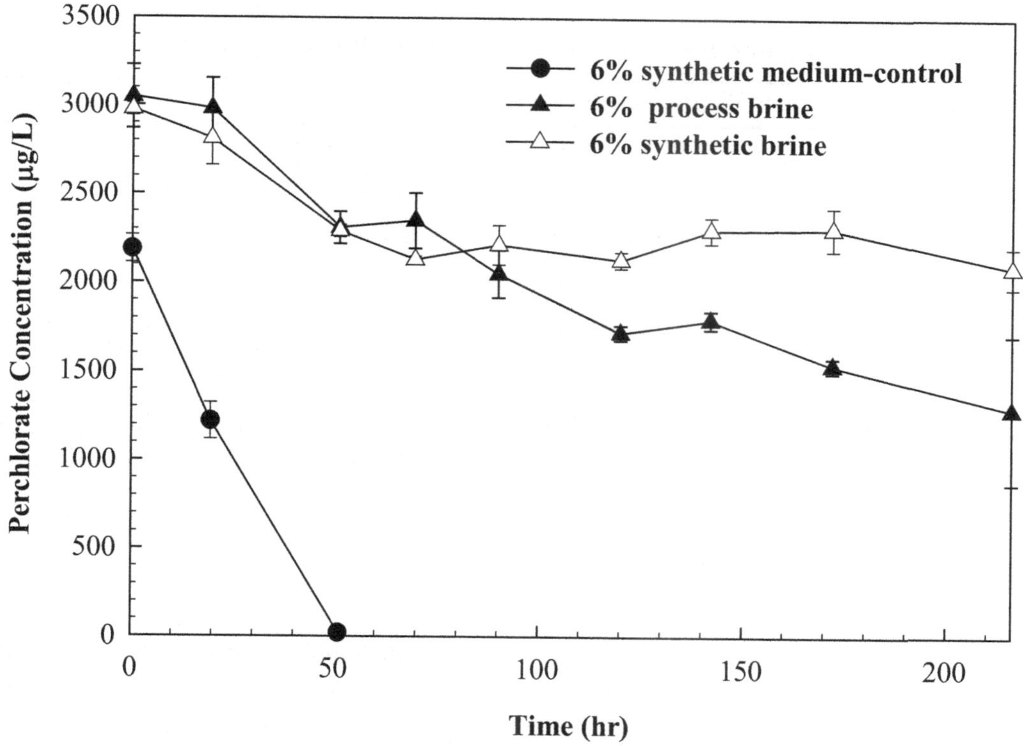

Figure 5.8 Brine effect on perchlorate degradation: 6 percent culture 3rd transfer

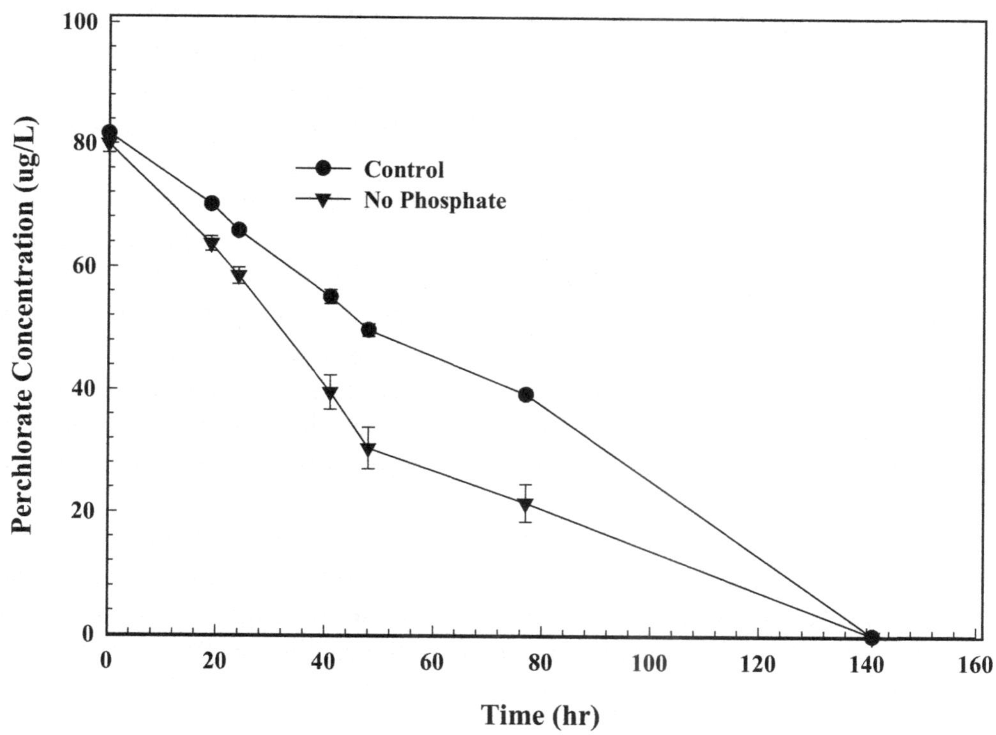

Figure 5.9 Effect of not including phosphate in the 6 percent synthetic medium on perchlorate removal

Table 5.6
Media changes used for the cation addition experiment

Test identification	Ingredient changes
Control	3 percent process brine (no cation addition)
Mg	3 percent process brine + 130 mg/L Mg
Ca	3 percent process brine + 40 mg/L Ca
K	3 percent process brine + 40 mg/L K
All (Mg+Ca+K)	3 percent process brine + 130 mg/L Mg, 40 mg/L Ca, 40 mg/L K

While phosphate is required for biological systems, this experiment showed that excessive levels may be counter productive. Additional tests were not performed to establish the maximum non-inhibitory concentration of phosphate.

Mixed Cations

One major difference between the synthetic medium and the brine (synthetic and process brine) was the presence of cations in the synthetic medium close to the concentrations in seawater. The seawater-adapted organisms in the cultures might have required these as growth factors or to balance the concentration of sodium (Na^+) ions in the brine.

3 Percent Culture

To evaluate the impact of calcium, magnesium and potassium on perchlorate degradation, individual cations were added to a series of 3 percent sub-cultures in spent process brine. The performance of these cultures were compared to unamended brine sub-cultures as well as sub-cultures that had received all three cations. Table 5.6 lists the five different conditions compared for this evaluation.

While differences between the perchlorate reduction performance of the sub-cultures was observed during the first transfer, they were more apparent after the second transfer. The addition of the individual cations increased the rate of perchlorate degradation over that of the unamended sub-cultures (see Figure 5.10). The sub-cultures receiving all three cations yielded the best performance.

In addition to perchlorate, the concentrations of nitrate and the major cations were monitored during these tests. The addition of the individual cations had a larger impact on the removal of nitrate than perchlorate. With the first transfer, obvious differences were already observed between the unamended brine sub-cultures as compared to those that received additional cations (see Figure 5.11).

Neither the potassium nor magnesium concentrations significantly changed while perchlorate was being reduced. The concentration of calcium, however, was significantly reduced after each addition. This result is not surprising considering the alkalinity of the brine was close to 2000 mg/L. With this alkalinity, a significant portion of the calcium would be expected to precipitate out of solution as calcium carbonate ($CaCO_3$).

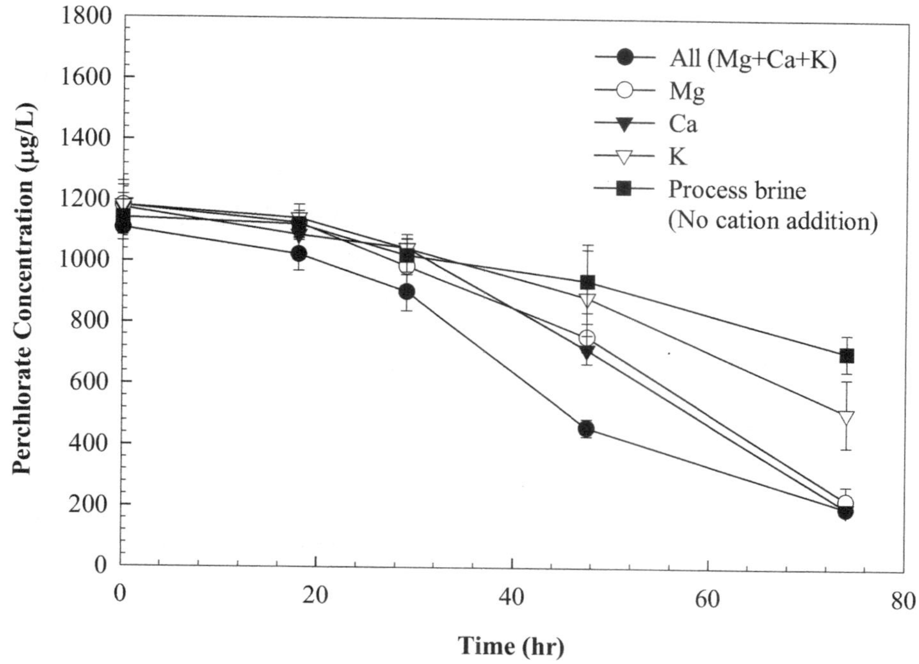

Figure 5.10 Effect of cations on perchlorate degradation: 3 percent culture 2nd feed

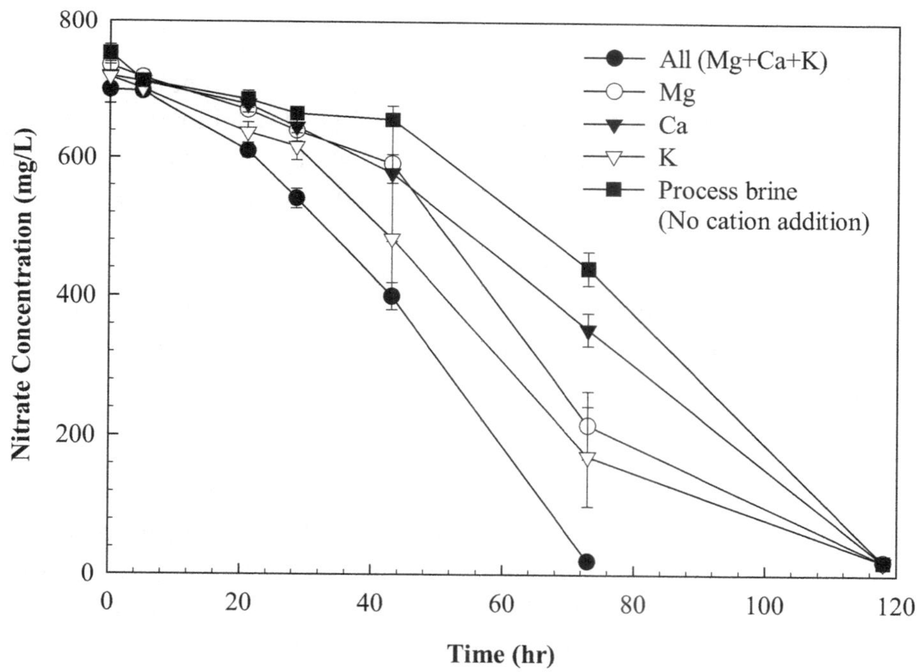

Figure 5.11 Effect of cations on nitrate degradation: 3 percent culture 1st feed

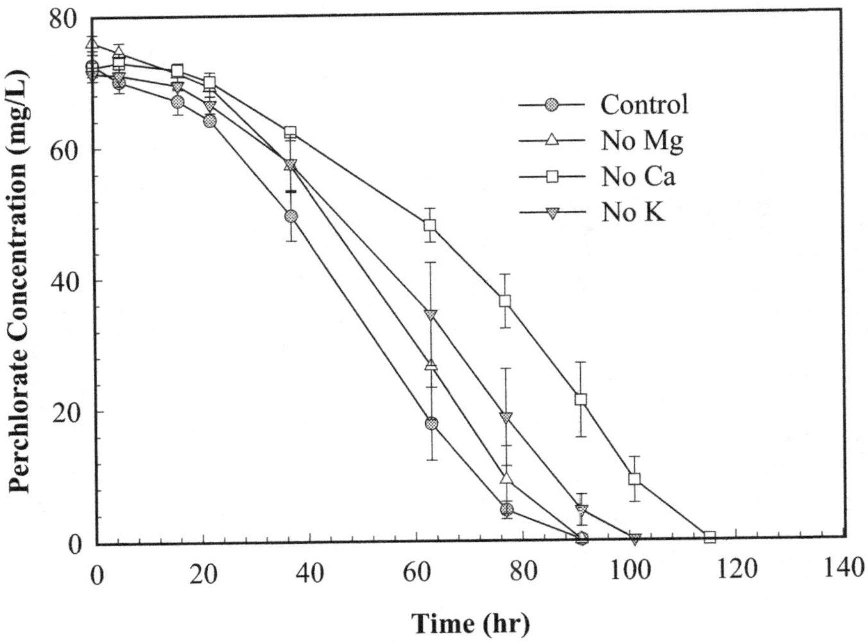

Figure 5.12 Perchlorate removal in media deficient of individual cations: 6 percent culture 2nd transfer

6 Percent Culture

Additional tests were performed with the 6 percent culture to help establish how these cations might affect perchlorate degradation. To avoid the loss of calcium through precipitation, this series of tests was prepared using the synthetic medium. As the synthetic medium contained each of the cations, they were selectively removed from each of the sets of sub-cultures. While no significant difference was observed between the sub-cultures after the first transfer, Figure 5.12 shows that the anticipated trend became apparent by the second transfer. These data revealed that perchlorate degradation slowed when any of the major cations were absent from the medium. The absence of calcium seemed to have the largest effect on perchlorate degradation.

Magnesium

While calcium seemed important for the reduction of perchlorate, its continued addition to a process brine would be costly as it would have to be added each brine treatment cycle to replenish the quantity that had precipitated. Additionally, the precipitate would need to be separated from the biomass (if possible) and removed from the bioreactor on a regular basis. If the precipitate was associated with the biomass, this would result in increased difficulties in keeping the biomass in suspension without imparting unnecessary shear to the culture.

In contrast to calcium, the magnesium concentration did not change during perchlorate reduction. Consequently, its addition would not be expected to suffer the same difficulties that calcium would. The following set of experiments was performed to determine if increasing the magnesium concentration would achieve the same effect as having calcium present. The experiments

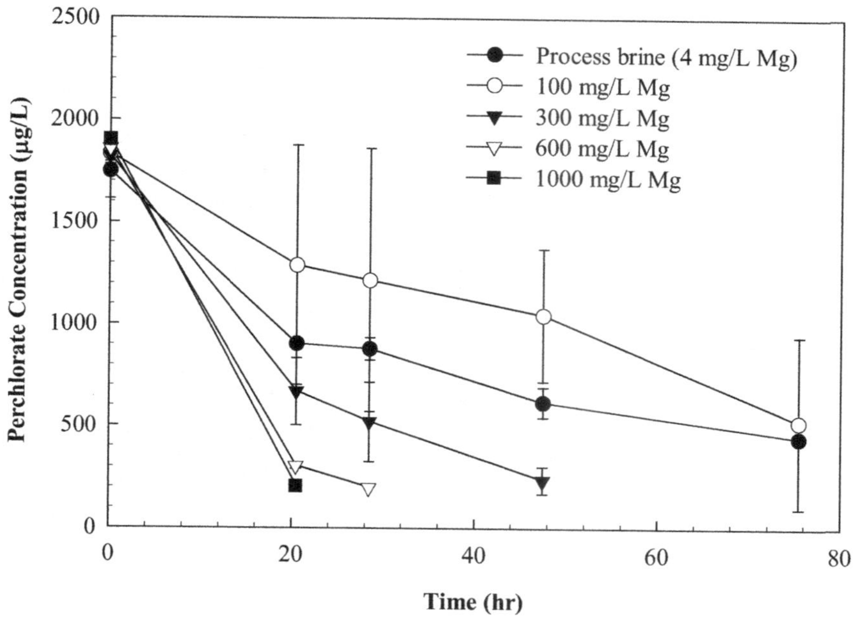

Figure 5.13 The effect of increasing concentrations of Mg^{2+} on perchlorate degradation: 3 percent NaCl 1st feed

were also meant to determine if magnesium was a suitable amendment to the brine to improve perchlorate reduction and what was a reasonable concentration range to consider. For this series of tests, a range of magnesium doses were spiked into 3 percent NaCl spent process brine. These amended brine samples were then inoculated with biomass collected from the 3 percent synthetic medium culture. The five different magnesium conditions were:

- Control (3 percent process brine – 4 mg/L Mg)
- 100 mg/L Mg addition
- 300 mg/L Mg addition
- 600 mg/L Mg addition
- 1000 mg/L Mg addition

As shown in Figure 5.13, immediate differences between the rates of perchlorate degradation were observed with the first transfer. These data show that in general, the rate of perchlorate degradation increased with increasing magnesium concentrations. The data for the 100 mg/L Mg concentration sub-cultures had significant scatter, likely due to inconsistencies with inoculation. By the third transfer, however, similar results were observed between the different sub-cultures of this condition, as shown in Figure 5.14. Additionally, the performance of the 100 mg/L Mg sub-cultures improved to the point that their rate of perchlorate degradation fell between the unamended process brine and the 300 mg/L Mg spiked systems.

During these tests the nitrate concentration was also monitored. Small differences were observed during the first transfer, as shown in Figure 5.15. It should be noted that during regular maintenance of the 3 percent NaCl culture in the synthetic medium, nitrate is not added. As a result, it took 2 to 3 days to reduce the nitrate. Once exposed to nitrate, however, the culture

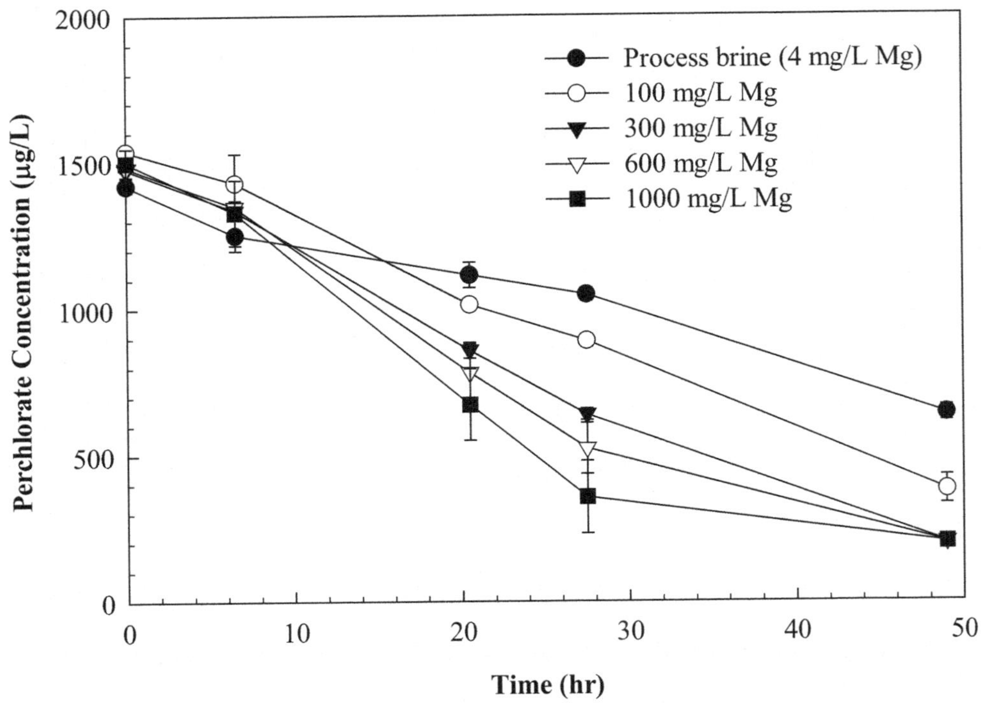

Figure 5.14 The effect of increasing concentrations of Mg^{2+} on perchlorate degradation: 3 percent NaCl 3rd feed

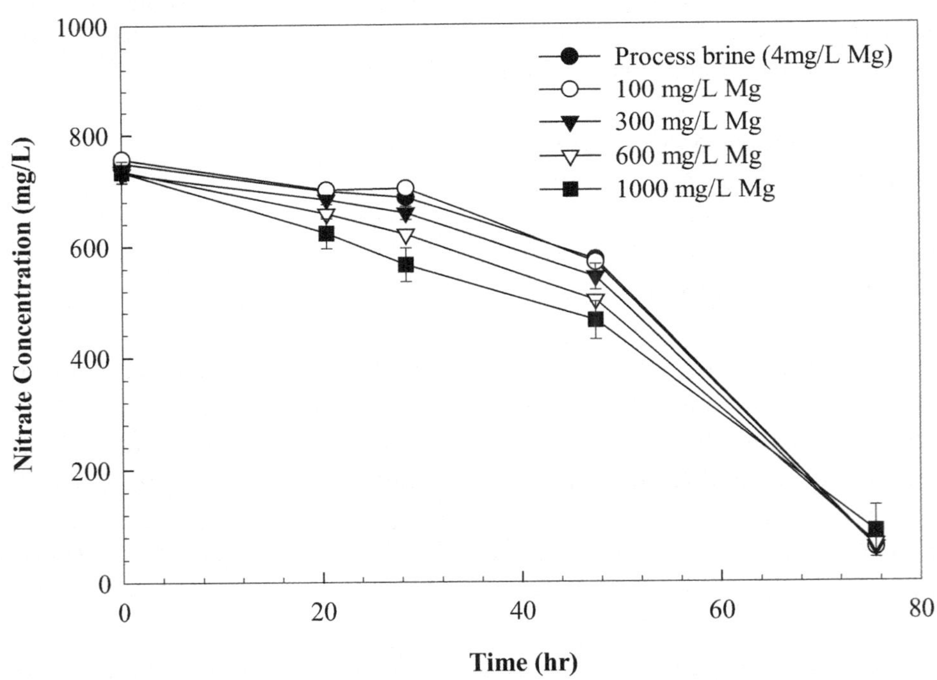

Figure 5.15 The effect of increasing concentrations of Mg^{2+} on nitrate degradation: 3 percent NaCl 1st feed

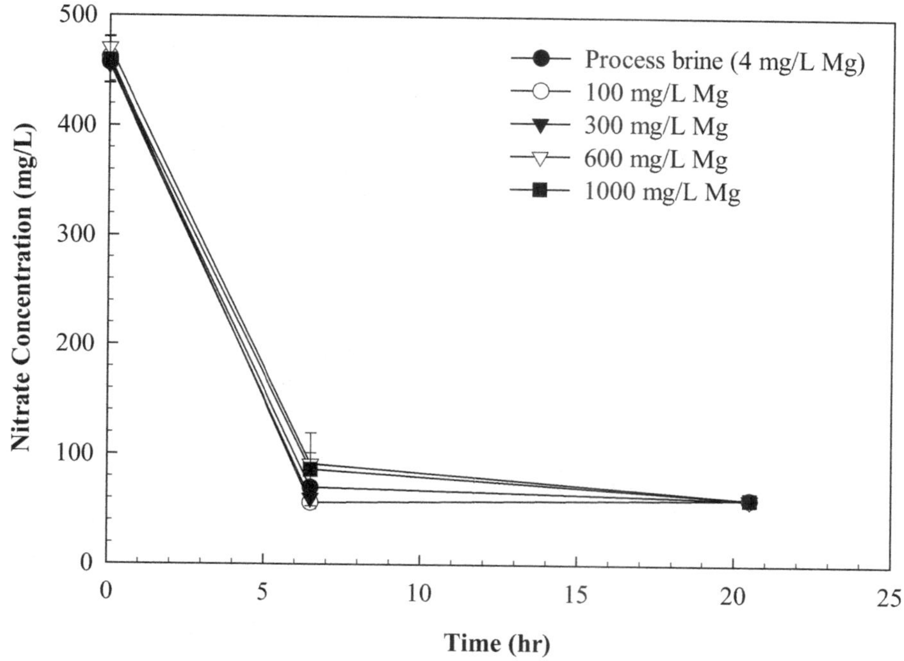

Figure 5.16 The effect of increasing concentrations of Mg^{2+} on nitrate degradation: 3 percent NaCl 3rd feed

quickly acclimated. It was not possible to distinguish between the nitrate degradation performance of the sub-cultures for the second transfer as all of the nitrate had been degraded by the time the first sample was collected (21 hours). Figure 5.16 shows that while samples were collected earlier during the third transfer (6.5 hours), essentially all of the nitrate had been degraded by this time so that no differences could be identified, if they existed.

Acetate

The culture requires an electron donor to degrade both the perchlorate and nitrate. The growth yield from carbon substrates in high salt conditions is usually lower than under non-saline conditions. Consequently, a larger amount of carbon source may be necessary under high salt conditions. Sub-cultures were prepared using the 6 percent process brine to determine the required amount of acetate to achieve perchlorate and nitrate removal from the ion exchange brine. Based on stoichiometry and the concentrations of perchlorate and nitrate present in the brine, 1.5 g/L of sodium acetate was required to remove all of the nitrate and perchlorate. The performance of the culture fed this amount was compared to two other series of cultures fed 2.25 and 3.0 g/L sodium acetate, 1.5 and 2.0 times the stoichiometric requirements, respectively. As shown in Figure 5.17, the excess acetate did not improve perchlorate reduction at high perchlorate concentrations. While not statistically significant with only a single set of tests in triplicate, it appears that a trend could be developing as the concentration of perchlorate decreased below approximately 1500 μg/L.

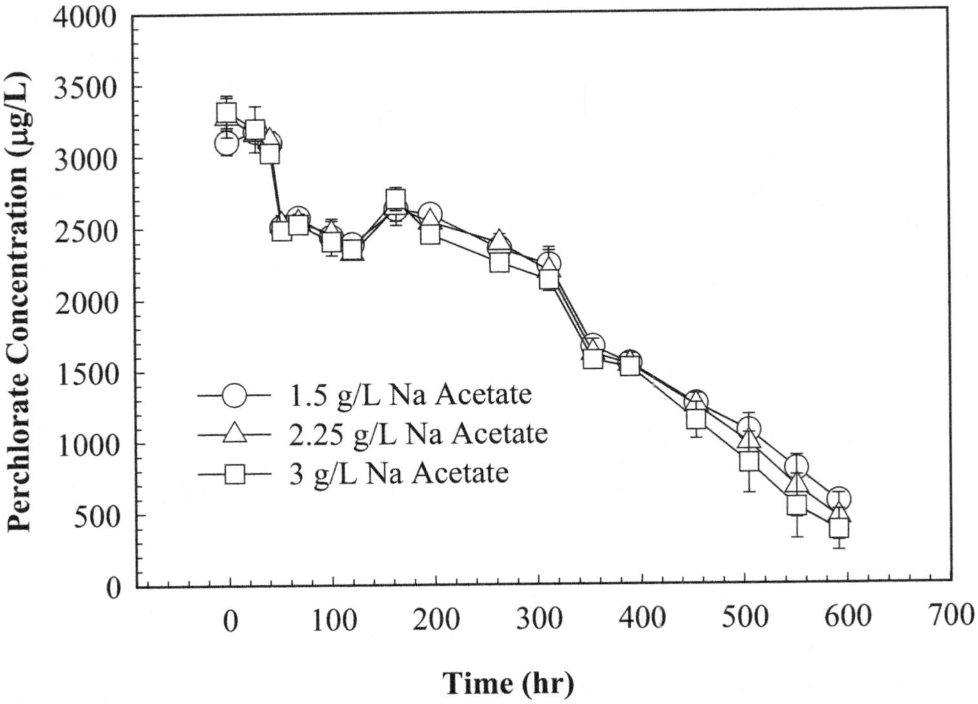

Figure 5.17 Effect of acetate on perchlorate reduction

IMPACTS ON CULTURE PERFORMANCE

Oxygen

In earlier work (Cang 2001), sodium sulfide (Na_2S) was determined to be necessary for continued culture performance, yet the observed effects of sulfide addition were inconsistent. This may have been due to the differences in exposure to oxygen through poor anaerobic culture techniques. Theoretically, when anaerobic conditions are effectively maintained through proper experimental techniques and sulfide is only present as an oxygen scavenger and reducing agent, the cultures exposure to molecular oxygen is limited resulting in minimal effects of sulfide. If experimental techniques were improper in maintaining anaerobic conditions, the opposite might be observed. To examine the effects of oxygen exposure and/or the addition of adding Na_2S (final concentration as 80.4 mg/L Na_2S-$9H_2O$) or FeS (~50 mg/L final concentration as FeS), triplicate cultures were inoculated from the 3 percent brine fed culture and exposed to various sulfide or oxygen exposure conditions (Figure 5.18).

The exposure of the culture to oxygen (improper anaerobic techniques) caused a complete cessation of perchlorate degradation. This condition was simulated by exposing the culture to air during the transfer and ensuring that the headspace (50 mL of the 150 mL serum bottles) was purged with ambient air before being sealed. The culture was then agitated to ensure entrainment of the headspace within the mixture. Even with the simultaneous addition of sulfide (either as soluble Na_2S or insoluble FeS), perchlorate reduction was completely inhibited within the 5-day incubation period. After the 5-day incubation period, additional attempts were performed to rescue the cultures in which bad technique was used. The bottles were flushed with N_2 gas, another

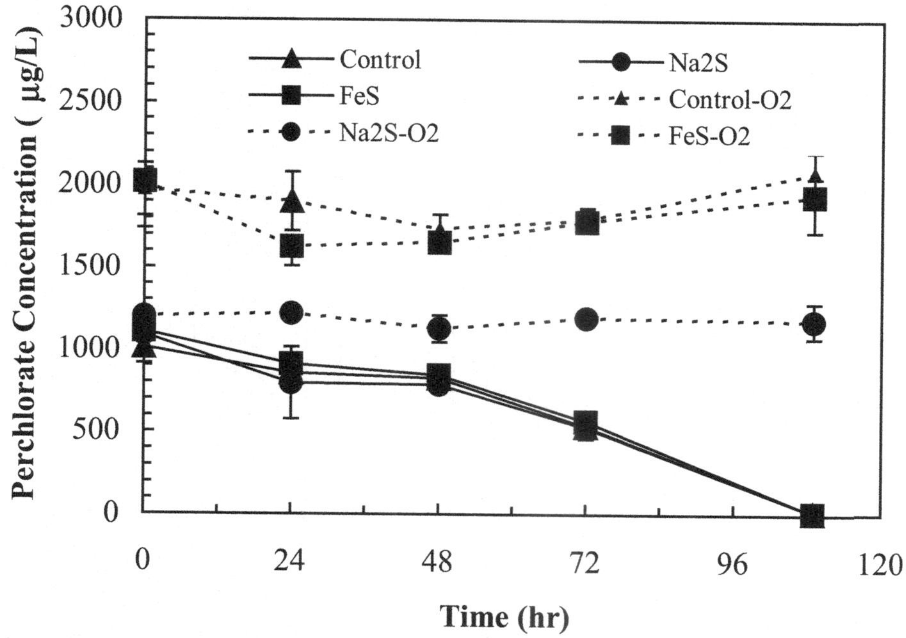

Figure 5.18 Impact of sulfide and oxygen exposure on perchlorate reduction

allotment of sodium sulfide was added to the cultures and additional acetate was added. The cultures were then able to degrade the perchlorate. For the cultures with which proper anaerobic techniques were employed, the addition of sulfide in either form did not improve perchlorate degradation.

pH

Throughout the laboratory experiments, the synthetic medium culture was usually maintained at a neutral pH (7), however, the pH of the brine used to convert the culture to a 3 percent brine feed was approximately 9. To assess the effect of pH on the performance of the 3 percent culture, five series of tests were established at initial pH values of 6.0, 7.0, 8.0, 9.0 and 10.0. The pH was adjusted using 2M HCl or 1M NaOH. These triplicate sub-cultures were also fed 1.5 times the stoichiometric acetate requirements and incubated while shaking (~100 rpm) at room temperature. The perchlorate concentrations were monitored over time, and the final pH was measured after the experiment. No attempt was made to adjust the pH during incubation.

The results, presented in Figure 5.19 and Figure 5.20, reveal that the optimal initial pH for perchlorate reduction was between pH 7 and 9, with a slightly faster degradation rate observed when the initial pH was 9. Outside of this range, however, the cultures were very sensitive to pH. At pH 10, only half of the perchlorate initially present was reduced in a two-week period. At pH 6, essentially complete inhibition was observed over a 10-day incubation period. It should be noted that the final pH of the pH 7, 8, and 9 cultures were 8.0, 8.7 and 9.0, respectively. These changes in pH were attributed to proton consumption during denitrification and potentially the organisms adjusting their environment to one they preferred. Results of these experiments showed that pH reduction to maintain a neutral pH is not necessary and may actually result in the slowing of perchlorate reduction. As a result, the optimal operational pH was found to be between 8 and 9,

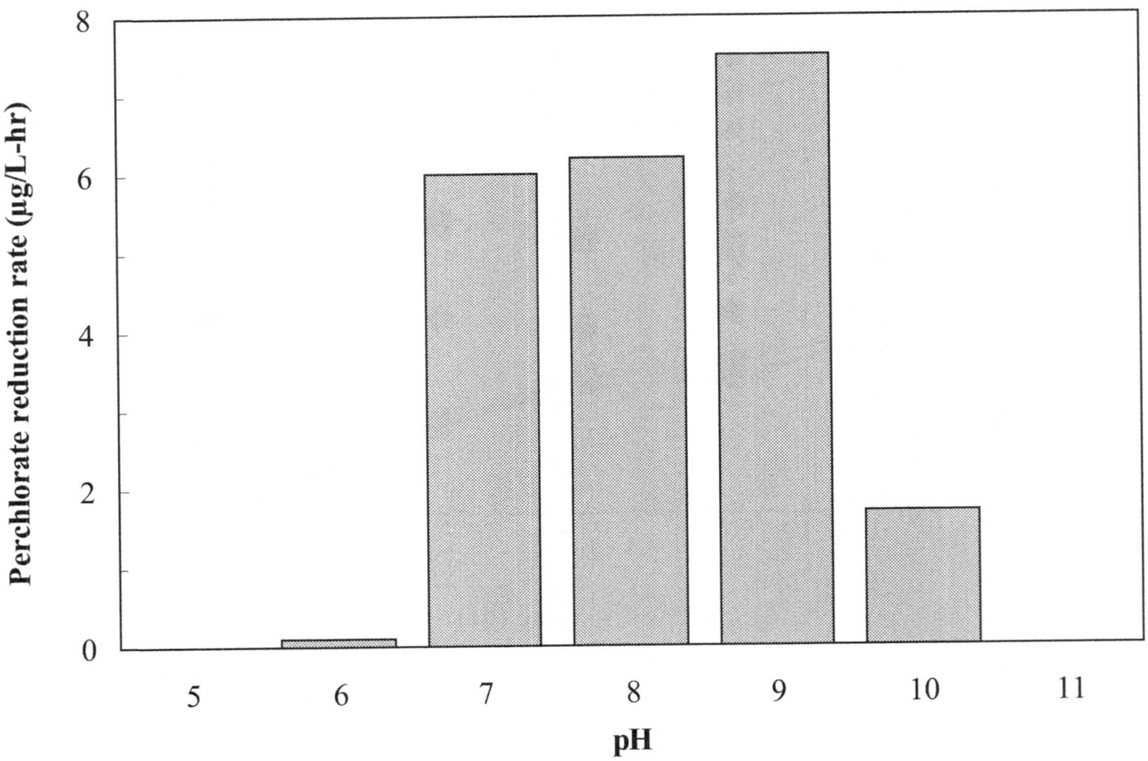

Figure 5.19 Effect of pH on rate of perchlorate degradation by the 3 percent culture

which was the typical pH observed during pilot-testing in the spent brine after regeneration of the exhausted resin.

PILOT-SCALE CULTURE DEVELOPMENT

Pilot Inoculum

In order to generate the inoculum for the pilot-scale testing, the seawater culture was grown up to a volume of 16 L and then acclimated to spent brine. To achieve this, approximately 3 L of synthetic seawater culture was transferred into 9 L of fresh-made synthetic seawater medium; approximately 120 g of original marine sediment inoculum was also added. After the perchlorate was reduced (9 days), the culture was spiked to 180 mg/L with fresh perchlorate. This was removed within 2 days and the culture was then fed an additional 4 L of brine to bring the total volume to 16 L and spiked to 800 mg/L perchlorate. The 800 mg/L of perchlorate was removed within 4 days. Figure 5.21 presents the perchlorate concentrations in the culture during this period.

The culture supernatant was then replaced with diluted brine three times using a SBR feed procedure. The first transfer was a 30 percent replacement. The second and third transfers were 40 and 50 percent replacements, respectively. In the second transfer, the perchlorate was spiked to 120 mg/L to encourage biomass growth. A perchlorate spike up to 110 mg/L was performed before the third replacement. During the third replacement the perchlorate concentration was

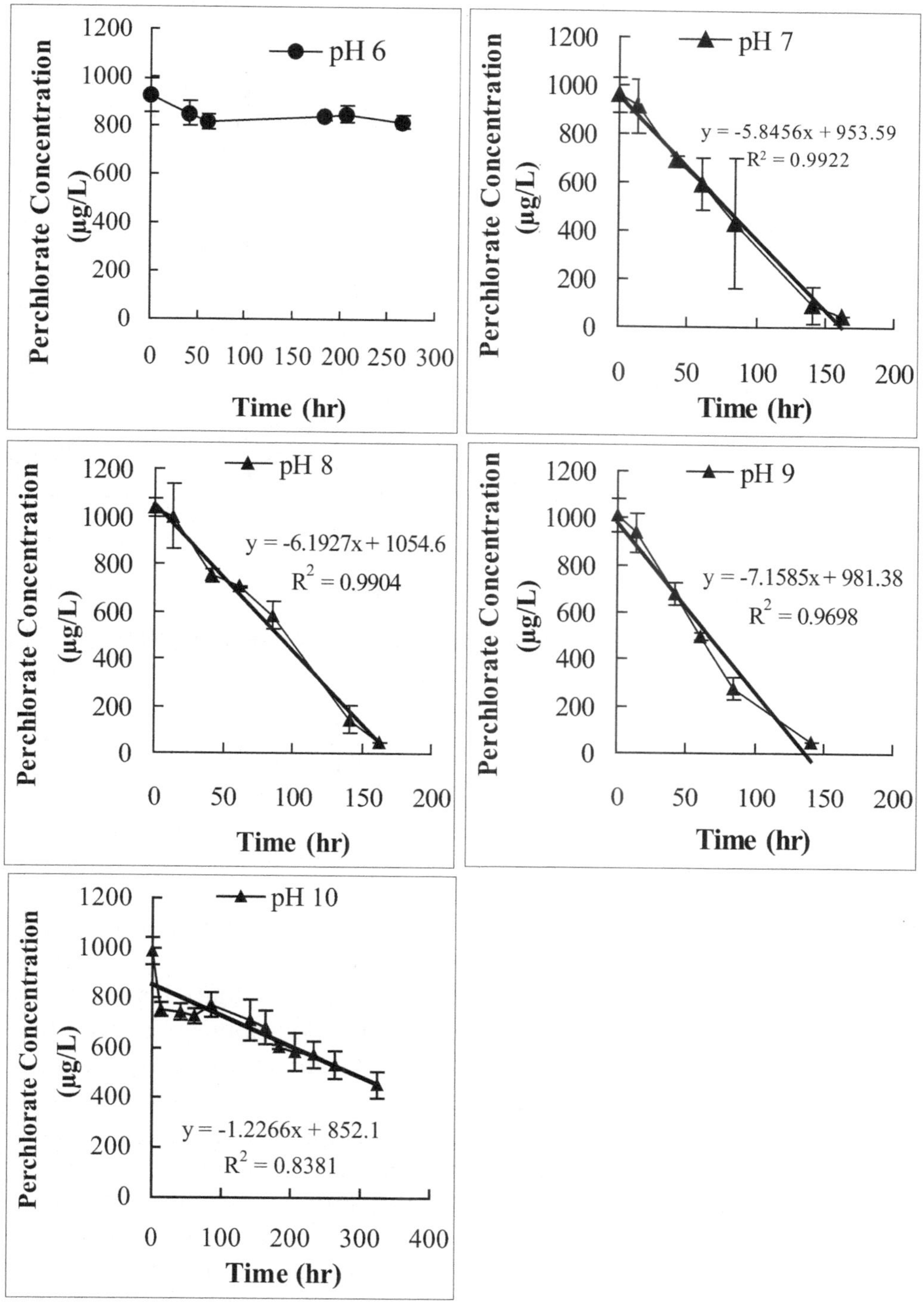

Figure 5.20 Perchlorate reduction in cultures initiated at different initial pH values

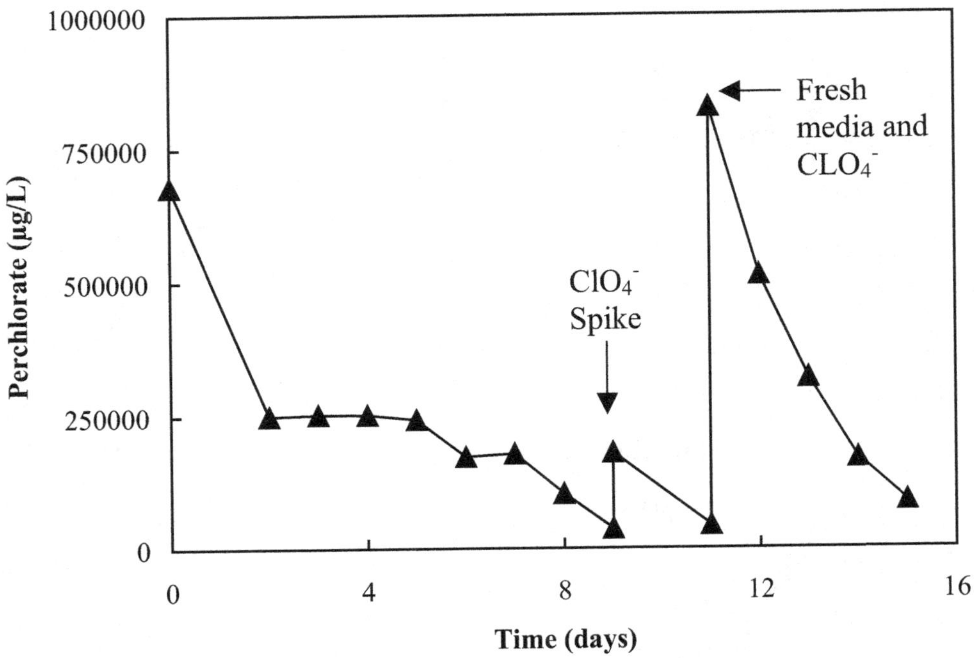

Figure 5.21 Growth of 3 percent synthetic seawater culture for pilot-scale use (3 to 16 L)

spiked up to 170 mg/L. Figure 5.22 graphically shows the performance of the culture during these transfers. This 16 L culture was then used to inoculate the pilot-scale 208 L (55 gal) reactor.

Culture Growth

At the pilot plant, the 16 L culture was transferred into the pilot reactor and 45 liters of 3 percent spent brine were added with 50 g of potassium acetate and 5.3 g of potassium perchlorate. Within the next 24 hours, the initial 54 mg/L perchlorate was reduced by 74 percent. Throughout the next week, additional brine was added daily and periodically spiked with perchlorate and the stoichiometric required amounts of acetate. Table 5.7 summarizes the development of the culture to the pilot-scale volume of approximately 208 L. As the culture volume was increased, complete reduction of perchlorate was observed within 24 hours. However, when the culture finally reached 200 L, perchlorate reduction slowed to 50 percent. After the first 60 percent transfer with fresh spent brine that had been spiked with perchlorate, reduction slowed even further to 20 percent in 24 hours. No extra nitrate was added during the culture growth other than the 600 mg/L originally available in the spent brine. Although perchlorate reduction appeared to slow during the first 60 percent transfer, complete nitrate reduction was consistently observed to occur with 10 hours.

Pilot Performance

The decreased performance of the culture at its full volume (208 L) was determined to be the result of low biomass (VSS of approximately 100 mg/L). Concurrent with the bench-scale investigation of the cultures divalent cation requirement, additional testing was conducted at pilot-scale in an attempt to restore culture performance. As shown in Table 5.8, the 3 percent

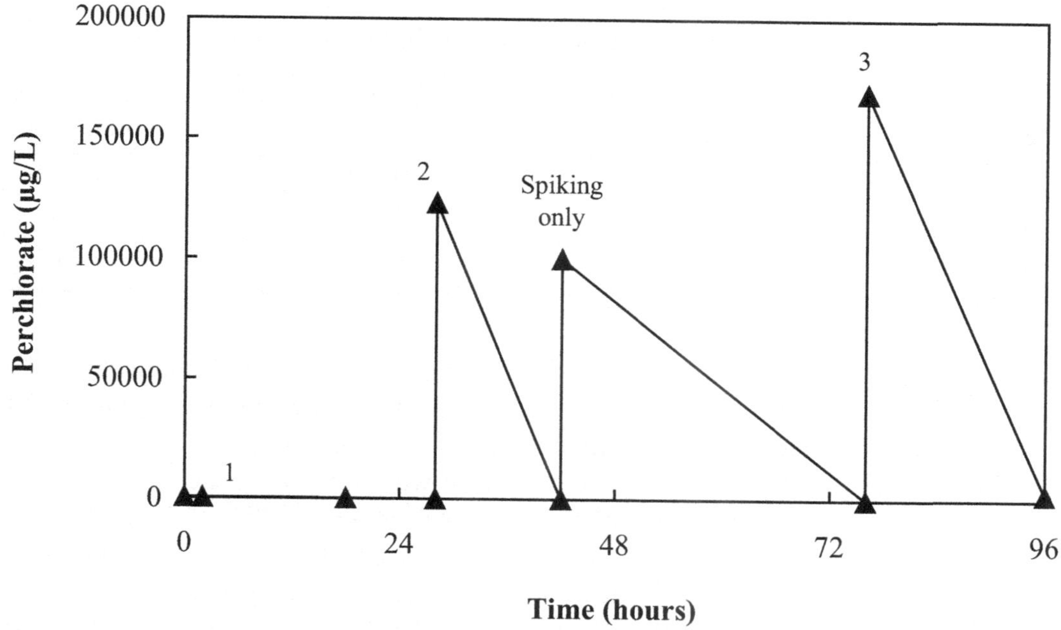

Figure 5.22 Conversion of pilot-scale culture to brine feed (3 percent NaCl)

Table 5.7
Summary of pilot culture growth

Final volume (L)	Brine added (L)	Brine transfer (%)	$C_o = ClO_4^-$ (mg/L)	ClO_4^- reduction in 24 hours (%)
60	40	—	54 *	74
60	—	—	156 *	81
80	20	—	9.2	100
100	20	—	41.2 *	100
120	20	—	1.6	100
120	—	—	1.5	100
140	20	—	2.2	100
200	60	—	0.78	50
200	—	60	45 *	20

* Spiked with perchlorate.

Table 5.8
Seawater versus pilot cation balance

Parameter	Typical seawater (mg/L)	Initial 3 percent brine (mg/L)	Amended 3 percent brine (mg/L)
Na^+	10,500	11,000	11,000
Ca^{2+}	400	65	65
Mg^{2+}	1270	12	800

brine generated at pilot-scale lacked significant amounts of calcium and magnesium, compared to typical seawater concentrations. As the bench-scale investigation was ongoing, the pilot-culture was initially amended with approximately 400 mg/L calcium.

As shown in Figure 5.23, perchlorate reduction was slow when the brine contained little to no calcium (No Ca). Based on the perchlorate removal rate of the "No Ca" condition, the time to reduce 90 percent of 1.5 mg/L perchlorate was estimated to be 19 days. Once approximately 400 mg/L calcium was added to the brine (Ca-1), culture performance dramatically improved and reduced perchlorate to below 200 μg/L in 2 days. Perchlorate was spiked daily along with additional calcium (to compensate for the quantity lost due to precipitation) over the next four days (Ca-2, 3, 4, and 5). For Ca-2 through Ca-4, 90 percent removals were achieved in 24 hours. On the fifth day (Ca-5), however, the culture performance once again suffered and the 1.5 mg/L perchlorate spike was reduced in approximately 3 days. In agreement with concurrent bench-scale tests, significant amounts of calcium were being lost to precipitation as $CaCO_3$ during each amendment creating a deficiency of soluble calcium available to the perchlorate reducing bacteria. To satisfy an observed divalent cation requirement to balance out the sodium ions, magnesium was used in further pilot-scale amendments, since it was not expected to be lost through precipitation. Following the Ca-5 spike, magnesium was amended to the 3 percent brine at 800 mg/L and culture performance was again restored (90 percent removal in 24 hours). The amendment of magnesium was advantageous because it was a successful substitute to calcium addition and less than 5 percent of the initially amended magnesium was lost to precipitation after each cycle/perchlorate spike. Throughout the remainder of the biological brine treatment testing, the magnesium concentration was maintained at 800 mg/L by periodic amendments of additional magnesium when needed.

PILOT BRINE TREATMENT AND RECYCLING

Using the operating conditions established during the baseline resin evaluation, a fresh batch of polyacrylic resin was exhausted for 425 BV at a flowrate of 1.7 gpm, and counter-current regenerated with 25 lbs/ft^3 using 3 percent NaCl brine. Spent brine was amended with acetic acid as an electron donor allowed to react. Treated brine from the SBR was then used to refill the regeneration tank and was amended with NaCl to bring the brine concentration back to 3 percent.

A total of 20 cycles were completed for this testing. Figure 5.24 compares the exhaustion profiles of perchlorate, nitrate, sulfate, chloride, and bicarbonate observed during selected runs (Run 1, 5, 10, 15, and 20). Effluent analysis for all other runs indicated similar performance. As seen in these anion profiles, a high degree of repeatability was observed during the course of biological brine treatment testing. While this consistency seemed to have indicated that recycling did

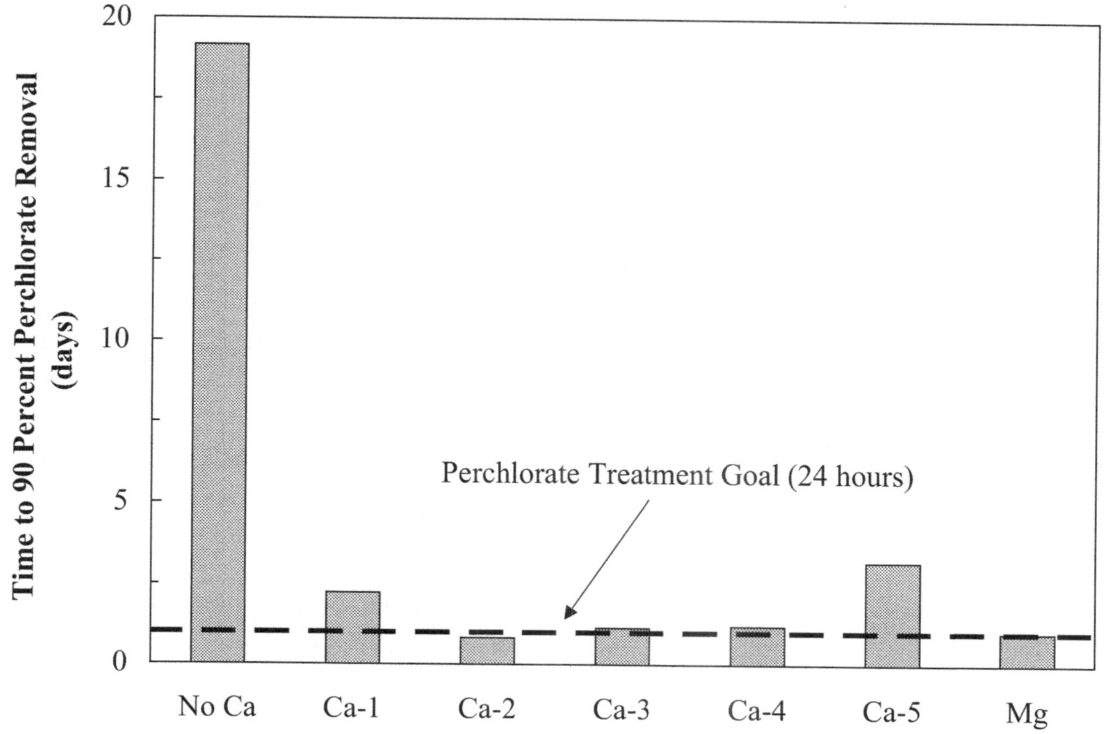

Figure 5.23 Effect of divalent cation addition on perchlorate reduction

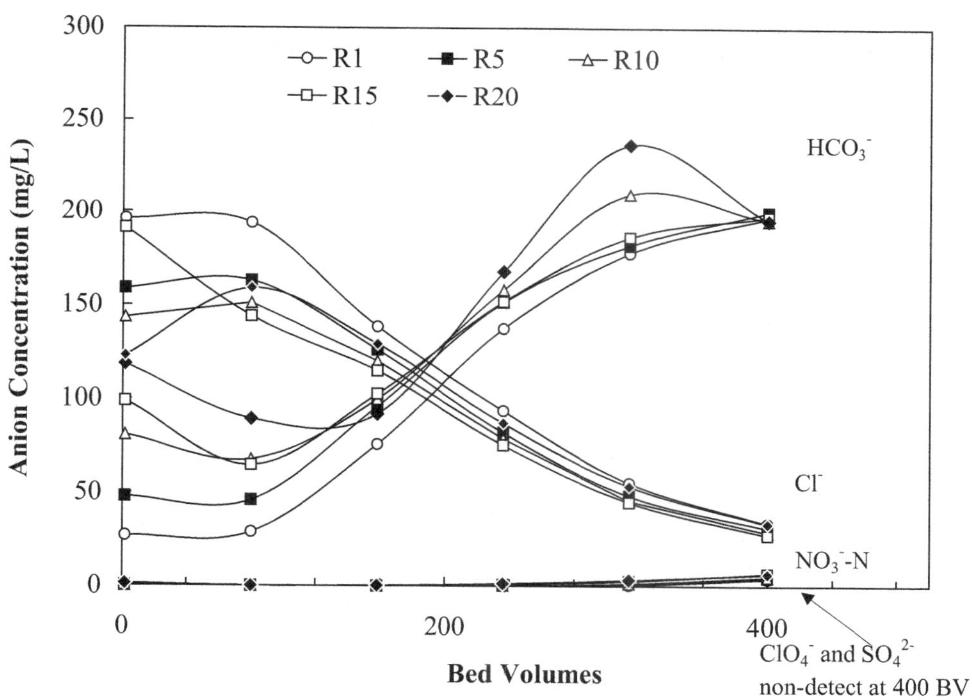

Figure 5.24 Comparison of exhaustion profiles during biological brine treatment (poly-acrylic resin)

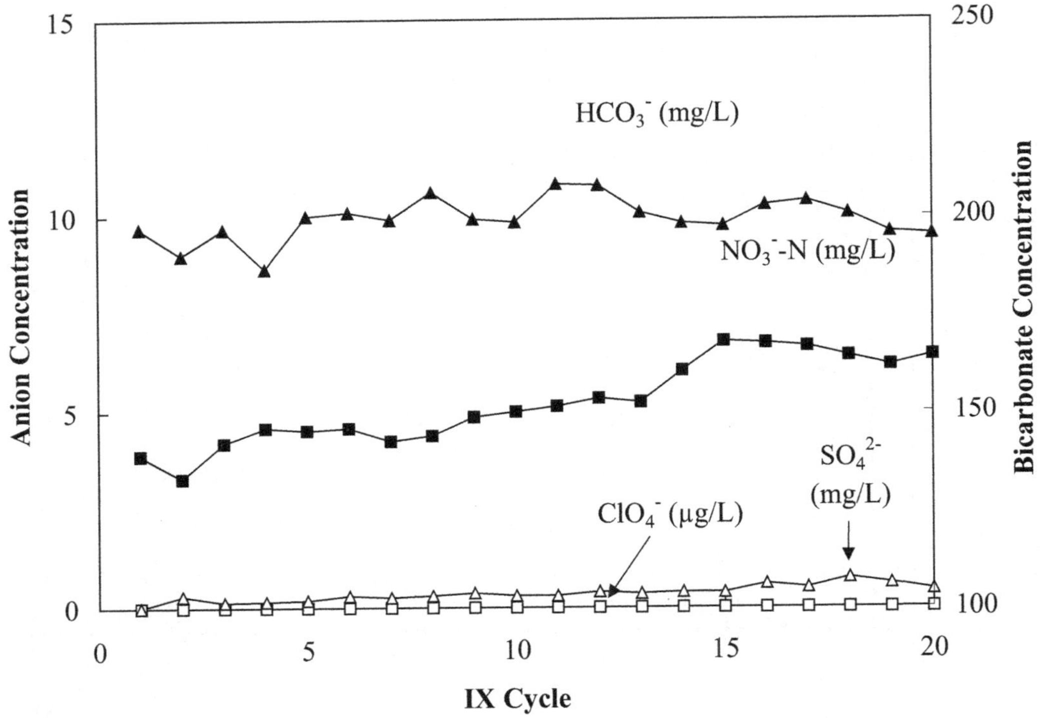

Figure 5.25 Effect of biological brine treatment and recycle on resin performance (poly-acrylic resin)

not impact bicarbonate and nitrate, further analysis of all of the runs revealed a general trend not detectable by comparing selected runs.

Minor differences between runs was expected due to the variability of the influent water quality and manual control of the regeneration process. As shown in Figure 5.25, the accumulation of selected anions (i.e., sulfate) in the brine and on the resin over the 20 cycles was reasoned to be accountable for the reduction in the number of available sites for nitrate (perchlorate) exchange and concomitant decrease in time to nitrate (perchlorate) breakthrough. Should the performance of the system continue to degrade to the point that operational goals be jeopardized, there would be several options that could be pursued to recover the performance of the system. The options that could be considered would include:

- Disposal of old brine and re-start with a new batch of regenerant brine, including:
  - Disposal to an ocean outfall,
  - Discharge to evaporative ponds, and subsequent disposal of dried salts to municipal land fill;
- Continuous discharge (partial or total) and amendment of the brine to limit the accumulation of sulfate,
  - Disposal as discussed above,
  - Reintroduce the discharged brine to the treated water. This would result in a final sulfate concentration equal to that of the untreated water, and a slight increase in TDS as a result of the additional NaCl addition,

- Treatment of the brine to remove sulfate,
  - Treatment by nanofiltration membranes would result in a large portion of the brine being ready for reuse, and a smaller fraction loaded with sulfate that must then be discharged as described above.

Counter-current regeneration was carried out at a flow rate of 0.37 gpm for 13.5 BV (salt loading rate of 25 lbs/ft^3) using a 3 percent NaCl solution. Prior to regeneration, the suspended biomass in the anaerobic SBR was allowed to settle for a minimum of one hour. A 65 percent transfer of the bioreactor supernatant was performed by pumping approximately 130 L of treated brine into the regeneration tank. During the transfer, unsettled biomass was removed using an inline 5-15 μg/L filtration process. After the contents of the SBR had been transferred, the brine was amended with NaCl to bring the brine concentration back to 3 percent and used for regeneration of the subsequent cycle.

Spent brine generated during the regeneration process was diverted directly to the SBR for treatment of perchlorate and nitrate during which the contents of the SBR were slowly mixed and continuously sparged with nitrogen to maintain an anoxic environment. Once the regeneration was complete, the spent brine acetate concentration in the SBR was raised to 1.5 times the stoichiometric requirement.

Biological reduction of nitrate and perchlorate to below the treatment goal was achieved for all 20 runs. Nitrate was quickly reduced to below analytical detection limits within 8 hours. The concomitant reduction of perchlorate, however, required approximately 30-40 hours to be reduced to below 200 μg/L by the end of the 20 cycles. Sulfate and bicarbonate were not removed by the biological brine treatment system. As a result, their respective concentrations in the brine increased with each consecutive regeneration. Figure 5.26 represents anions present in the composite brine following a regeneration procedure. Perchlorate and nitrate concentrations remained consistent for all 20 runs averaging 1100 μg/L and 90 mg-N/L, respectively. The slight variability in the increasing sulfate and bicarbonate trend was associated with the variable influent concentrations of respective anions.

Detailed regeneration profiles of nitrate and perchlorate for selected runs (Runs 1, 5, 10, 15, and 20), as seen in Figure 5.27 (a) and Figure 5.27 (b), were nearly identical to the results found during baseline testing. Complete removal nitrate and perchlorate (below 200 μg/L) in the spent brine ensured that the treated brine could successfully be reused numerous times without a significant loss in resin performance. In addition, the accumulation of sulfate and bicarbonate in the treated brine had no effect on the regeneration performance of perchlorate and nitrate when recycled. In Figure 5.27 (c) and Figure 5.27 (d), the effect of the accumulation of bicarbonate and sulfate in the spent brine and on the resin can be seen in regeneration profiles for selected runs (Run 1, 5, 10, 15, and 20).

WATER QUALITY

A major sampling campaign was performed to characterize the water quality during biological brine treatment and recycling. Particular attention was given to the brine water quality before (spent brine) and after biological brine treatment.

Acetic acid was amended to the spent brine as the electron donor before being processed by the brine culture. To ensure that donor was available in excess for the brine culture and verify that the excess donor did not affect the treated water quality during exhaustion of the resin, acetate

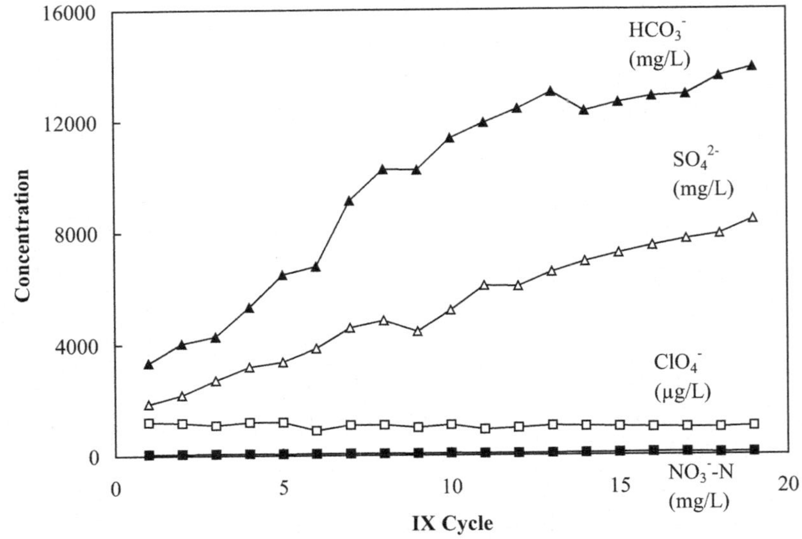

Figure 5.26 Composite recycled brine anions prior to biological brine treatment (poly-acrylic resin)

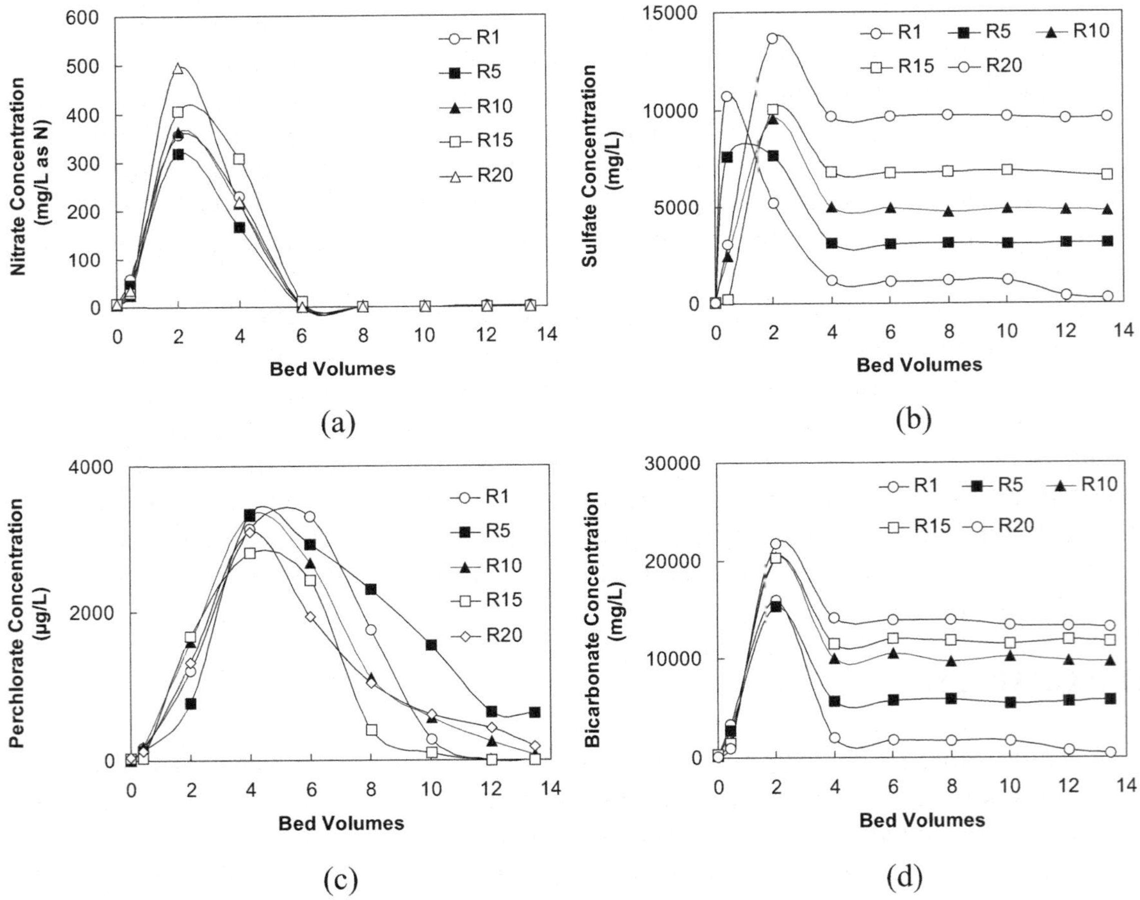

Figure 5.27 Composite recycled brine anions prior to biological brine treatment (poly-acrylic resin)

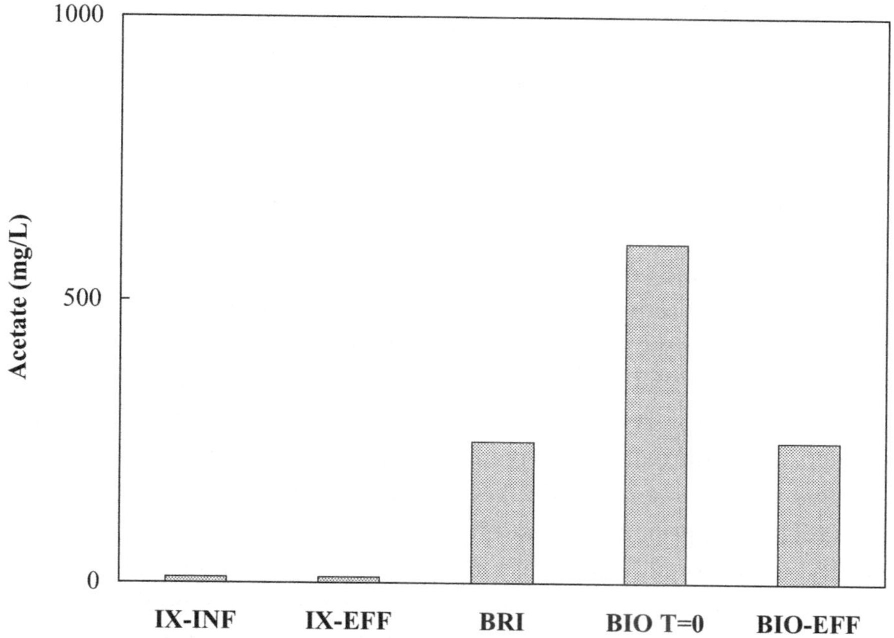

Figure 5.28 Profile of acetate through the treatment process

Table 5.9
Biologically treated brine water quality parameters with noticeable increase

Parameter	Run 1	Run 30
Sulfate	1600 mg/L	8300 mg/L
Bicarbonate	4000 mg/L	17500 mg/L
Uranium	175	870 μg/L
Fluoride	1.5	3.8 mg/L

was monitored throughout the treatment process. As shown in Figure 5.28, the brine culture solution was amended to approximately 600 mg/L acetate, corresponding to the stoichiometric requirement based on perchlorate and nitrate in the spent brine and a targeted residual. The acetate residual was maintained at 200 mg/L. It is important to note that no acetate was detected in the treated effluent from the ion-exchange column at any time during exhaustion, despite the fact that the resin was regenerated with recycled brine containing 200 to 300 mg/L acetate. This verified that the 3 BV of rinse water used was sufficient to displace any residual acetate.

Table 5.9 summarizes selected water quality parameters that had a noticeable increase over the course to the 20 cycle biological treatment and recycle experiment. Sulfate and bicarbonate concentration, as expected, accumulated in the treated brine as the brine culture only biologically removed perchlorate and nitrate. Other parameters included uranium and fluoride. These elements, measured in low amounts in the raw water, accumulated after each regeneration and were not removed by the brine culture.

OPERATION AND MAINTENANCE

Operation and maintenance (O&M) requirements were minimal for the successful operation of the biological brine treatment system. The brine culture was able to successfully reduce perchlorate and nitrate in the complex brine matrix containing high TDS salts and a pH of 8 to 9. Two critical operational requirements were identified during experimental testing: mixing and maintaining an anoxic environment.

Mixing is critical to ensure brine culture cells are suspended to promote biomass production and efficient perchlorate reduction kinetics. The mixing, however, must not be so aggressive that undue shear is placed on the biomass as this may inhibit the growth of colonies. In addition, care must be taken to ensure that the mixing mechanism and operation does not compromise the anoxic nature of the bioreactor. As experienced in the early stages of pilot testing, operational failures occurred when a standard motor driven shaft type mixer was used. First, the some of the mixer's components (i.e. set screws, collars, fittings) were unable to withstand the corrosive nature of the brine. The components, incidentally, were not constructed of stainless steel and were damaged. Secondly, operational failures associated with oxygen poisoning occurred as the closed reactor was compromised at the point between the reactor housing and the mixing shaft. With dissolved oxygen present in the bioreactor, culture performance suffered dramatically and required nearly 5 days to achieve reduction of perchlorate.

Several modifications were applied to the bioreactor mixing system to minimize the potential for oxygen to be introduced to the reactor. These modifications included utilizing alternative motor mounting configurations, sealants, and suitable materials of construction for all reactor components in direct contact with the brine. Additionally, nitrogen was slowly bubbled into the reactor to maintain a nitrogen atmosphere over the culture with internal reactor pressure slightly elevated above ambient atmospheric pressure.

MICROBIOLOGICAL ANALYSIS OF CULTURES

Gram stains and attempts at pure culture microbiology were conducted with the 3 percent and 6 percent laboratory cultures. Originally, nine pure cultures were obtained from the 3 percent cultures (fed either synthetic medium or process brine collected from the pilot ion exchange system) developed in the laboratory. These pure cultures were all capable of reducing nitrate but not perchlorate under anaerobic conditions. Subsequent attempts to isolate pure cultures from the active cultures yielded fewer colonies. The following material focuses on the organisms that were identified in the pilot scale culture.

Well-mixed culture samples were collected for microbiological characterization tests at the end of the first (R1) and last (R20) recycle of the brine to track the changes in the microecology. The samples were characterized by perchlorate-reducer enumeration, nitrate-reducer enumeration, aerobic Plate Count Agar (PCA) plating, Gram's stain, and VSS measurements. These samples were also fixed and stored for future DAPI stain and probing.

Enumeration of Organisms

The most probable number (MPN) technique was used to estimate perchlorate- and nitrate reducer counts in the pilot culture samples. The synthetic medium was used for growth of perchlorate reducing organisms in the MPN tubes. A series dilution, 10^0 through 10^{-5} of the pilot culture,

Table 5.10
Counts of specific organisms in the pilot culture samples

Group enumerated	R1	R20
Perchlorate-Reducers*	230	2800
Nitrate-Reducers*	$>10^7$	1.7×10^{10}
Aerobic Heterotrophs†	7.1×10^6	$\sim 10^6$

* (MPN/100 mL).
† CFU.

was prepared by serial dilution of the sample into phosphate buffer solution (PBS). The PBS contained 30 g/L NaCl, 1.24 g/L Na_2HPO_4, and 1.8 g/L $NaH_2PO_4 \cdot H_2O$. After serial dilution, 1 mL of the sample (or dilution) was inoculated to MPN tubes, which contained 9 mL of synthetic medium. These samples were incubated in the dark, at room temperature for 21 days. Tubes were interpreted as positive when at least 70 percent of the initial allotment of perchlorate was degraded. The data in Table 5.10 reveal that an increased in the number of perchlorate-reducers in the pilot-scale culture was observed during the recycle tests.

An enumeration of nitrate-reducing organisms was also performed using MPN techniques. This test detects the ability of organisms to reduce nitrate to nitrite or nitrogen gas in nitrate reduction broth. The procedure of this test was similar to MPN test described before. A serial dilution of 10^0 through 10^{-9} of the pilot culture into PBS was used to inoculate MPN tubes containing 9 mL of 3 percent NaCl nitrate reduction broth (Difco) spiked with $MgCL_2 \cdot 6H_2O$ (final concentration ~11 g/L), $CaCl_2 \cdot 2H_2O$ (final concentration ~1.4 g/L) and KCl (final concentration ~0.72 g/L). These were incubated in the dark, at room temperature for 21 days. A positive tube was one in which gas was collected in a central inverted tube and the media in the tube tested negative for nitrate and nitrate. The results of the tests are also presented in Table 5.10.

For the first sample tested, the dilution series was only carried out to the 10^{-5} dilution. Positive reactions were observed in all of the dilutions therefore the correct dilution was missed, the results for this are presented as the minimum number that could be in the sample. The numbers of nitrate reducing-organisms also increased during pilot plant operation. These results were consistent with the results of VSS measurement, which increased from 41 mg/L in the first sample to 234 mg/L in the last sample.

The plate count method was applied to enumerate and purify the organisms that grew on the aerobic Plate Count Agar (PCA) plates. For the first sample, growth was observed on PCA plates 2 days after inoculation. Well-formed colonies were observed after 4 days. At least 4 different colonies appeared on the plates. Descriptions of these colonies including genetic identification will be presented in the next section. There were fewer total heterotrophs than nitrate reducing organisms (Table 5.10). Table 5.11 presents a breakdown of the numbers of each colony type and their representation in the total population.

No new colonies appeared on the PCA plates for the R20 sample. No enumerations were performed because when the plates were put in the refrigerator, after 4 days incubation but before counting, vigorous growth and spreading of colonies B, C and D occurred. This indicated that the bacteria forming colonies B, C, and D could continue to grow at low temperature (4°C). The countable dilutions would have been the same as for the first sample, so it is assumed that the numbers of aerobic heterotrophs did not increase markedly.

Table 5.11
Numbers of aerobic heterotrophs in the first sample

Colony	CFU/mL	% of population
A	7×10^5	10
B	1.5×10^6	21
C	3×10^5	4
D	4.6×10^6	65
Overall	7.1×10^6	100

Results of Colony Purification and Identification

Pure cultures were obtained by selecting individual isolates from the PCA plates from the first sample and successively streaking them onto fresh PCA plates. These procedures resulted in the isolation of 6 colony types in pure culture from the pilot scale culture, namely, A-1, A-2, A-3, B, C, and D. Colonies A-1, A-2 and A-3 were isolated from the colony that is documented as colony A in Table 5.11. Colony A-1 and A-3 had different colony morphologies, but similar cellular morphology as shown by Gram's stain (Figure 5.29). Colony A-2 had a similar colony morphology to Colony A-1 but showed a different cellular morphology as shown by Gram's stain. Colony B, C, and D had individual Gram's stain and colony morphologies. Discussions of the identification and importance of these organisms is documented below. These cultures were also tested for their ability to reduce perchlorate or nitrate in liquid culture (synthetic media). Interestingly, none of these isolates were able to reduce perchlorate in liquid medium even though all of the cultures were capable of nitrate reduction.

A Series Colonies

The organisms named Colony A were large, dry, creamy colored round colony with undulate edge and a wrinkled surface. The size varied from 1/16 to 1/8 inches. The initial Gram's stains of the Colony A after purification attempts suggested it was not a pure culture even though the plates only contained one colony morphology, further attempts purify this colony yielded three colonies, in which one appeared different macroscopically (smaller, moister and not wrinkled) and one that although it appeared the same macroscopically it had different cellular morphologies (Figure 5.29).

Colony A-2 was not sent to MIDI labs for identification because its colony morphology was very close to A-1. The difference of the appearance of the Gram Stain between A-1 and A-2 may be due to the cells being in different growth stages when sampled. The 16srRNA sequences of colony A-1 and A-3 were identical, although their morphologies on PCA plates were different. Furthermore, there was only 0.1 percent genetic difference (GD) between the sequences of the organisms in Colonies A-1and A-3 and *Pseudomonas stutzeri or Pseudomonas fragi*. Figure 5.30 shows the genetic differences in the 16sRNA sequence for these three organisms. In general, a match less than 1.0 % GD may be considered a species level match. *P. stutzeri* is a denitrifier and generally found in soil and water (Palleroni 1984). *P. fragi* is a temperature sensitive organism incapable of growth at 37°C, but has been known to cause food spoilage in refrigerated foods, especially milk. The genus *Pseudomonas* belongs to the Gammaproteobacteria group.

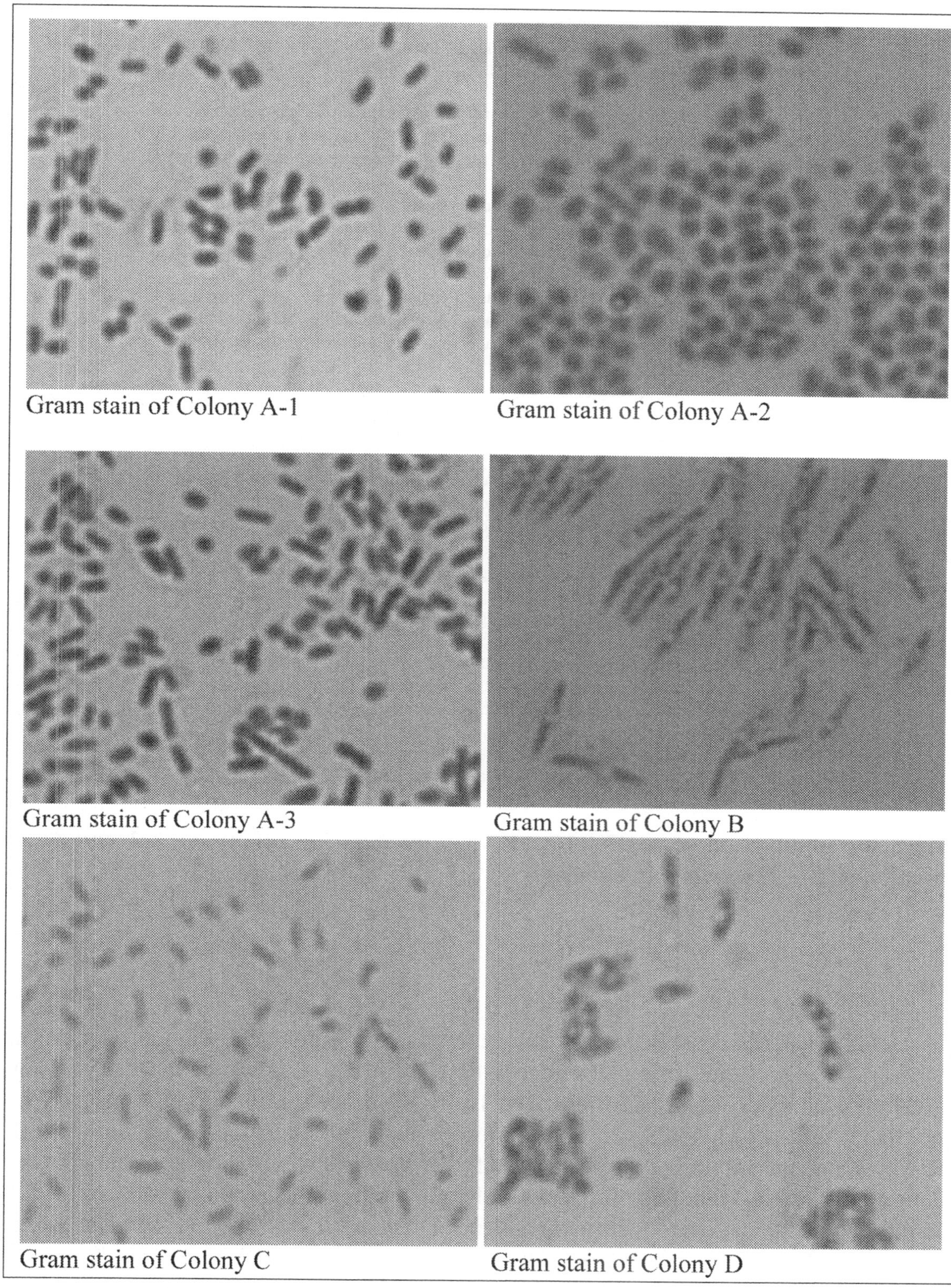

Gram stain of Colony A-1

Gram stain of Colony A-2

Gram stain of Colony A-3

Gram stain of Colony B

Gram stain of Colony C

Gram stain of Colony D

Figure 5.29 Gram stains of colonies A-1, 2, 3, B, C, and D

```
Colony A:     6   agtttgatcctggctcagattgaacgctggcggcaggcctaacacatgcaagtcgagcgg 65
P. fragi:     1   agtttgatcctggctcagattgaacgctggcggcaggcctaacacatgcaagtcgagcgg 60
P. stutzeri:  2   -gtttgatcctggctcagattgaacgctggcggcaggcctaacacatgcaagtcgagcgg 60

Colony A:     66  atgagtggagcttgctccatgattcagcggcggacgggtgagtaatgcctaggaatctgc 125
P. fragi:     61  atgagtggagcttgctccatgattcagcggcggacgggtgagtaatgcctaggaatctgc 120
P. stutzeri:  61  atgagtggagcttgctccatgattcagcggcggacgggtgagtaatgcctaggaatctgc 120

Colony A:     126 ctggtagtgggggacaacg t ttcgaaaggaacgctaataccgcatacgtcctacgggaga 185
P. fragi:     121 ctggtagtgggggacaacg a ttcgaaaggaacgctaataccgcatacgtcctacgggaga 180
P. stutzeri:  121 ctggtagtgggggacaacg t ttcgaaaggaacgctaataccgcatacgtcctacgggaga 180

Colony A:     186 aagtgggggatcttc r gacctcacgctatcagatgagcctaggtcggattagctagttgg 245
P. fragi:     181 aagtgggggatcttc g gacctcacgctatcagatgagcctaggtcggattagctagttgg 240
P. stutzeri:  181 aagtgggggatcttc g gacctcacgctatcagatgagcctaggtcggattagctagttgg 240

Colony A:     246 t gaggtaaaggctcaccaaggcgacgatccgtaactggtctgagaggatgatcagtcaca 305
P. fragi:     241 t gaggtaaaggctcaccaaggcgacgatccgtaactggtctgagaggatgatcagtcaca 300
P. stutzeri:  241 c gaggtaaaggctcaccaaggcgacgatccgtaactggtctgagaggatgatcagtcaca 300

Colony A:     306 ctggaactgagacacggtccagactcctacgggaggcagcagtggggaatattggacaat 365
P. fragi:     301 ctggaactgagacacggtccagactcctacgggaggcagcagtggggaatattggacaat 360
P. stutzeri:  301 ctggaactgagacacggtccagactcctacgggaggcagcagtggggaatattggacaat 360

Colony A:     366 gggcgaaagcctgatccagccatgccgcgtgtgtgaagaaggtcttcggattgtaaagca 425
P. fragi:     361 gggcgaaagcctgatccagccatgccgcgtgtgtgaagaaggtcttcggattgtaaagca 420
P. stutzeri:  361 gggcgaaagcctgatccagccatgccgcgtgtgtgaagaaggtcttcggattgtaaagca 420

Colony A:     426 ctttaagttgggaggaagggcagtaagttaataccttgctgttttgacgttaccaacaga 485
P. fragi:     421 ctttaagttgggaggaagggcagtaagttaataccttgctgttttgacgttaccaacaga 480
P. stutzeri:  421 ctttaagttgggaggaagggcagtaagttaataccttgctgttttgacgttaccgacaga 480

Colony A:     486 ataagcaccggctaacttcgtgccagcagccgcggta 522
P. fragi:     481 ataagcaccggctaacttcgtgccagcagccgcggta 517
P. stutzeri:  481 ataagcaccggctaacttcgtgccagcagccgcggta 517
```

Figure 5.30 Comparison of partial 16s rRNA sequences for colony A (A-1 and A-3 were identical) and two *Pseudomonas* species

Colony B

Colony B can be described as a creamy color, circular colony with smooth edge and cracking surface. Its size varied from 1/64 inches to 1/32 inches. Gram's stain revealed long thin pointed rods with intracellular inclusions. The 16s rRNA sequence of the organisms in Colony B did not match any known species of bacteria using the MIDI lab genetic database or GenBank. There was 96 percent similarity with a species of *Azospirillum*, which was not a match for species, but indicate that it may belong in this Genus, belonging to the Alphaproteobacteria group. Colony B shared many characteristics with those given for *Azospirillum* (*Bergey's Manual*; Krieg and Dobereiner 1984). For example, when cultured in PCA agar, Colony B grew as slightly curved rods and straight rods and many of the cells had pointed ends. These characteristics were also found when *Azospirilla* were cultured in complex media such as MPSS broth (Krieg and Dobereiner, 1984). Colony B was also tested to be able to reduce nitrate and some species or strains of

Azospirillum were also denitrifiers. The other organisms that Colony B was similar to included unidentified environmental isolates from an ultradeep African gold mine or deep basalt aquifers of Washington State (Baker et al. 2003). No references could be found to any species of *Azospirillum* as a pathogen.

Colony C

Colony C can be described as a cream colored, circular colony with smooth edge and surface, size approximately 1/32 inches. The Gram stain revealed it as a Gram-negative rod. The 16s rRNA sequence of the organisms in Colony C matched a few known species of Marinobacter using the MIDI lab genetic database and GenBank. There was only 0.19 percent genetic difference (GD) between colony C and *Marinobacter hydrocarbonoclasticus* using the MIDI lab database. Few matches were found using the GenBank database. One match was Marinobacter R2, which was isolated from an Antarctic deep-sea sediment (Lin et al. 2002).

Colony D

Colony D can be described as a shiny opaque cream colored pinpoint colony with smooth edge and surface. The Gram's stain revealed that it is a Gram variable rod with large intracellular inclusions. The 16s rRNA sequence of colony D did not match to any known species of bacteria using the MIDI lab genetic database or GenBank. The 16s rRNA sequence similarity was 97 percent between colony D and the marine *Agrobacterium stellulatum*. However, *Agrobacterium stellulatum* forms star shaped clusters and colony D did not. No other information was available. The large intracellular inclusions could be poly-β-hydroxybutyrate (PHB), a known bacterial carbon storage polymer.

Gram Stain Results and Pure Culture Characteristics Summary

Gram's stains were performed on the laboratory 3 percent, laboratory 6 percent and pilot-scale 3 percent cultures throughout the research period. General observations of more than 50 Gram stains of the various cultures were that the laboratory cultures showed a great deal more clumping of cells than the pilot scale culture, and that all of the cultures continued to evolving with time.

Generally, six different kinds of bacteria were observed in both batch and pilot cultures under a microscope:

1. Gram negative, vibrio
2. Rod
3. Double rod
4. Long, thin, flexible rod
5. Cocci
6. Gram variable rod

Changes in population compositions were found even for the same culture at different times. This indicated that these mixed cultures were not at steady state.

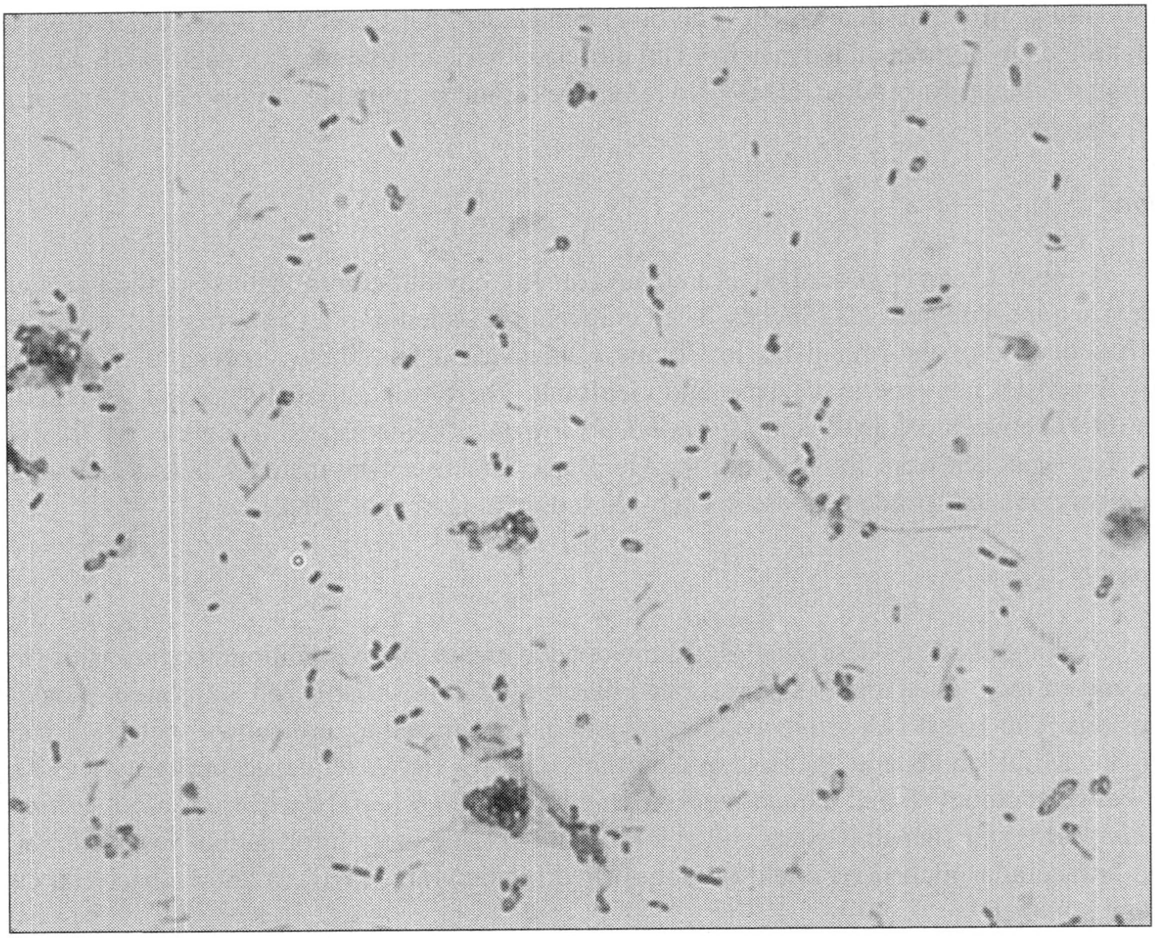

Figure 5.31 Photomicrograph of Gram stained sample of the pilot culture (oil immersion 1000× magnification)

General observations of the first sample from the pilot scale culture showed numerous large and small non-specific particles of unknown material. There was a possibility that some organisms were masked on or in these large particles. There were few cells, including Gram (-) rods; vibrio; and Gram variable rods with large internal accumulations. Less diversity of cells was observed in the last pilot-scale sample indicating that the culture was almost completely overtaken by the Gram variable rods with large internal accumulations, presumed to be internal carbon storage polymers. Figure 5.31 presents one of the photomicrographs of a Gram stain of a sample of the pilot scale culture.

Table 5.12 summarizes the results of the Gram stains performed on the cultures scored by abundance of each cellular morphology. The Gram negative medium sized rods, double rods, and Gram negative coccus may all be the same microorganism at different stages in its life cycle, but they do appear as distinct morphologies in the Gram's stains. These could all be either the *Pseudomonad* or the *Marinobacter*. These organisms were more prominent in cultures that were fed nitrate, but were also found in large numbers in perchlorate-only cultures.

Table 5.12
Summary of Gram stain observations

ID#	Description	3 percent synthetic medium culture		Culture from MPN test*	Pilot culture (3 percent)		6 percent synthetic medium culture		Possibly related to pure culture
		July, 2002	June, 2003		Recycle 1	Recycle 20	July, 2002	June, 2003	
1	Gram (-) vibrio	+++	+	+++	+	++	+++	+++	
2	Gram (-) rods (medium size)	+++	+++	+	+	++	+	+++	A1, C
3	Gram (-) double rods	+++	+++	+	+	+		++	A1, C
4	Gram (-) long, thin, flexible rods	+	+	+		+	+	+	B
5	Gram variable rod	+	+		+	+++	+++	+	D
6	Gram (-) cocci	+						+	A1, A2, C

* The culture was from one of the MPN tubes (10^0 dilution) for perchlorate-reducer enumeration for the pilot culture (recycle 20).
"+" the + symbol is used to indicate relative numbers of cells in the sample, the more + symbols the more cells seen.

The Gram variable rod definitely had a strong resemblance to pure culture D. It was also more dominant in the original sample of the 6 percent culture (salt tolerance) and the pilot scale culture. It almost disappeared from the 6 percent culture after one year of incubation in the absence of nitrate, but dominated most cultures fed nitrate.

The cells with the vibrio (comma shape) morphology were the only ones that would grow on aerobic or anaerobic agar. These were dominant in cultures fed only perchlorate and were present in all cultures that actively degraded perchlorate. Since attempts to grow this culture on solid agar failed, liquid culture dilution to extinction methods were employed. A 10 percent inoculum from one of the MPN tubes from the brine culture perchlorate-reducing enumerations was used to inoculate another liquid media tube. This was then allowed to degrade perchlorate and was transferred to a new tube twice more. A third transfer was unsuccessful. The second transfer has been saved for future studies.

Further investigation is required to allow the continued purification of the vibrio bacteria and characterization of nitrate reducing pure cultures. Once accomplished, the kinetics of perchlorate (nitrate) reduction and growth of the cultures can be modeled in a competition-based model. Additionally, a well-defined reactor seed culture would be developed.

BIOLOGICAL BRINE TREATMENT CONCLUSIONS

Results of the biological brine treatment system demonstrated that the process was able to consistently reduce the perchlorate and nitrate concentrations in the spent brine (0.5 N / 3 percent NaCl) to below the treatment goals (ClO_4^- < 200 µg/L and NO_3^- < 0.5 mg-N/L) for 20 (re)cycles. Sulfate and bicarbonate concentration increased to 8,500 mg/L and 14,000 mg/L, respectively, and did not affect the effluent water quality after each recycle. However, an increasing trend was observed with nitrate ion concentration in the finished water quality at the end of exhaustion. This clearly illustrated a slow decrease in the capacity of the resin for nitrate as various anions accumulated in the recycled brine and on the resin.

The accumulation of non-targeted anions in the recycled brine could be controlled by several methods including: (1) periodically wasting recycled brine and replacement with a fresh batch of brine, (2) periodically wasting a small portion of the recycled brine (and subsequently amending it with virgin sweet brine) to limit the accumulation of these other anions, or (3) further processing of the treated brine to remove some of these accumulating anions (i.e. passing the treated brine through a NF membrane).

Successful operation of the biological brine treatment system, however, requires careful monitoring to identify and prevent the following water quality, operational, and O&M issues:

- Although, sulfate and bicarbonate accumulated in the treated and recycled brine without deleterious effects, uranium accumulated from 175 µg/L to 8700 µg/L. This accumulation of heavy metals, radiologicals, or other constituents depend on water quality and may pose disposal and reuse issues.
- Mixing is a critical operational component of the system, which is required to ensure the suspension of biomass and to promote continued biomass growth. Mixing, however, must not be so aggressive that it physically disrupts or inhibits the growth of colonies.
- Oxygen is inhibitory to the salt-tolerant perchlorate-reducing microorganisms. Care must be taken to insure that an anoxic environment is maintained during treatment of

perchlorate-laden brine and that dissolved oxygen poisoning is minimized, especially, during brine replacements and after resin regeneration. The anoxic environment can be maintained by using a nitrogen sparge and/or the addition of sufficient sulfide which promotes a reducing environment for the microorganisms.

- The balance between the monovalent and divalent cations is necessary for rapid biological ClO_4^- reduction (patent pending).
- Due to the corrosive nature of brine, robust and compatible materials of construction should be appropriately utilized.
- The operation of the biological brine treatment could be conducted by the same plant operator that would normally monitor operation of an ion exchange plant. It is not necessary for the operator be a skilled microbiologist.
- Operating the system using redundant reactors would minimize biological system recovery time. The use of a redundant reactor design ensures that if a single system/reactor were to fail, several backup and concurrently running cultures/reactors would be available to eliminate or minimize the down time required to re-acclimate and grow a new culture. Additionally, the individual reactor volumes would be smaller and more manageable if multiple redundant reactors were to be used. In the event that all reactors fail, an alternative means of brine management (i.e., virgin brine regeneration and brine disposal) would be required to maintain the production perchlorate-free water while the biological brine treatment system is service offline. Recovery of the biological brine treatment system would necessitate some or all of the following tasks: cleanout of systems, obtain a new inoculum, and culture regrowth and acclimation to perchlorate-laden brine.

It is important to note that the biological brine treatment system pilot tested utilized a 3 percent NaCl culture. Current research efforts have been focused on further developing the brine culture to adapt to brine strengths up to 9 percent. Too date, results are promising and additional pilot testing at higher salt strengths will be performed to demonstrate performance. The realized benefit of operating the brine treatment system with an undiluted brine would be a decrease in water consumption and waste, and a smaller footprint for the SBR reactors thereby, reducing capital and operating costs (see Chapter 8).

CHAPTER 6
CHEMICAL BRINE TREATMENT

Several existing perchlorate treatment systems currently use Calgon Carbon Corporations ISEP® continuous ion exchange system. The La Puente Valley County Water District, the field-testing site, utilizes a 2,500 gpm ISEP® unit to remove nitrate and perchlorate. This conventional ion exchange process generates spent ion-exchange brine that must be disposed of in a dedicated brine line. Calgon has introduced the ISEP+™ system to the market. The "+" represents a catalytic destruct process used to destroy perchlorate and nitrate in spent ion-exchange brine. Alone, the catalytic destruct process is known as the perchlorate and nitrate destruction module (PNDM). Calgon provided a pilot-scale PNDM unit to be evaluated.

PNDM PILOT TESTING

Using the operating conditions established during the baseline resin evaluation, a fresh batch of polyacrylic resin was exhausted for 425 BV at a flowrate of 1.7 gpm, and regenerated with 25 lbs/ft^3 using a 6 percent NaCl brine. Spent brine from the regeneration was then dosed with ammonia and fed to the PNDM. Ammonia doses were based on the concentrations of perchlorate and nitrate present in the composite spent brine. Approximately 6 hours was required to treat the 17 gallons (66 L) of spent brine generated from the regeneration processes. Treated brine from the PNDM was used to refill the regeneration tank and amended with NaCl, if necessary, to bring the NaCl concentration back up to 6 percent.

Figure 6.1 compares the exhaustion profiles of perchlorate, nitrate, sulfate, chloride, and bicarbonate observed during selected runs (Run 1, 10, 20, and 30). Effluent analysis for all other runs indicated similar performance. As seen in these anion profiles, a high degree of repeatability was observed during the course of chemical brine treatment testing. While this consistency seemed to have indicated that recycling did not impact bicarbonate and nitrate, further analysis of all of the runs revealed a general trend not detectable by comparing selected runs.

While minor differences between runs was expected due to the variability of the influent water quality and the manual control of the regeneration process, the increasing trend in the final nitrate concentration seen in Figure 6.2 could not be attributed to these effects. The accumulation of other anions (i.e. sulfate) in the brine and on the resin over the 30 cycles, however, could account for a reduction in the number of available sites for nitrate (perchlorate) exchange and concomitant decrease in time to nitrate (perchlorate) breakthrough. Should the performance of the system continue to degrade to the point that operational goals may be jeopardized, the same options that would be utilized in the biological brine treatment system could be pursued to recover the performance of the system. The options that could be considered would include:

- Disposal of old brine and re-start with a new batch of regenerant brine, including:
  - Disposal to an ocean outfall,
  - Discharge to evaporative ponds, and subsequent disposal of dried salts to municipal land fill;

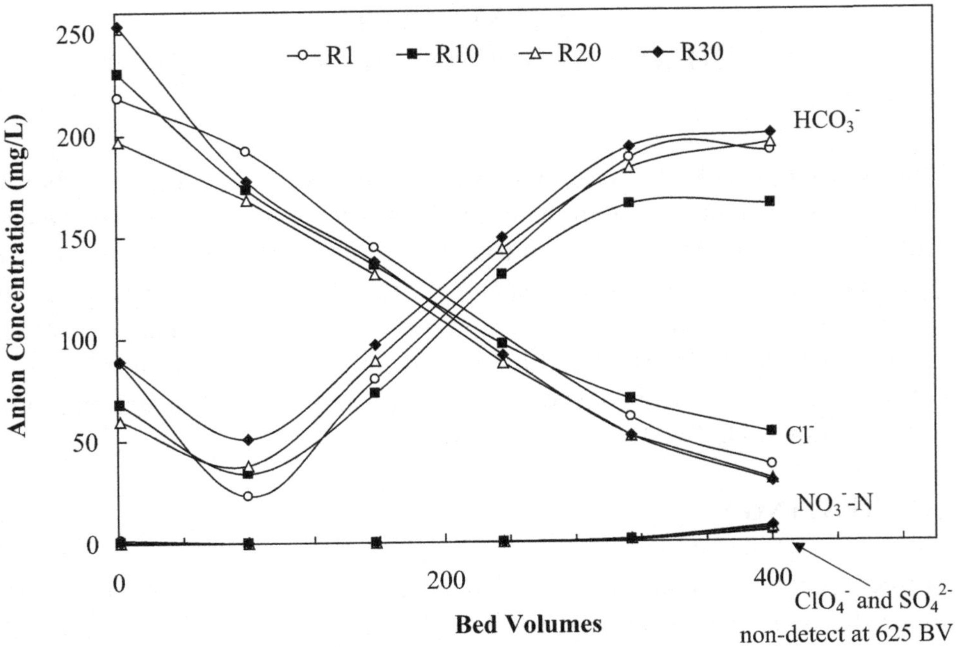

Figure 6.1 Comparison of exhaustion profiles during chemical brine treatment (polyacrylic resin)

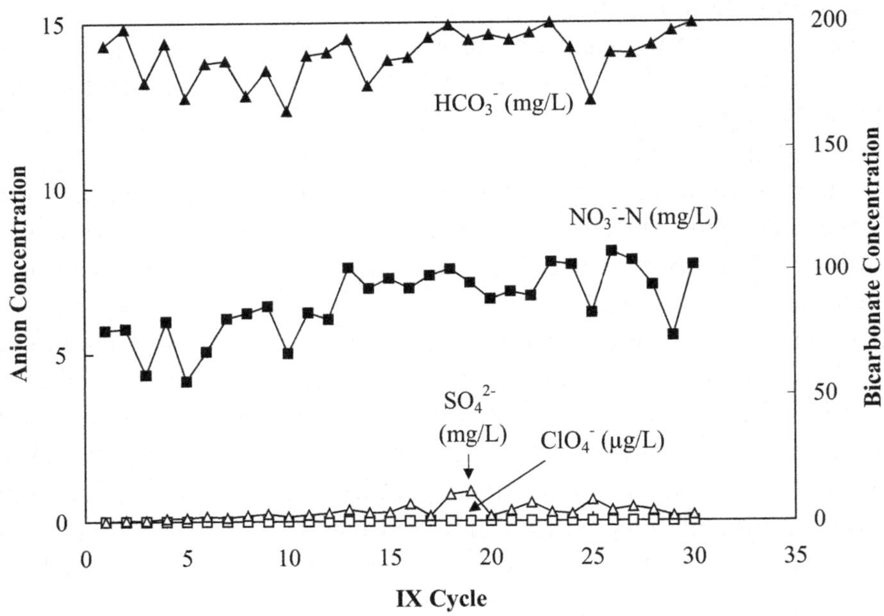

Figure 6.2 Effect of chemical brine treatment and recycle on resin performance (polyacrylic resin)

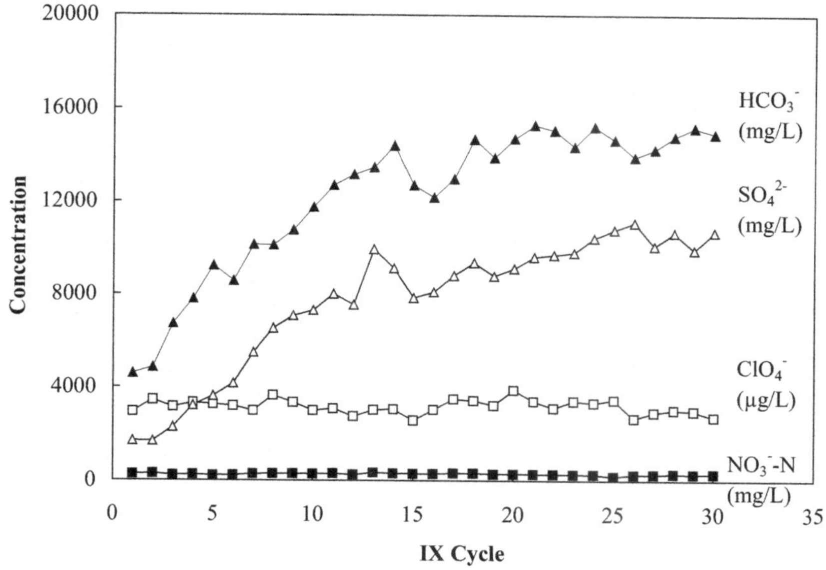

Figure 6.3 Composite recycled brine anions prior to chemical brine treatment (polyacrylic resin)

- Continuous discharge (partial or total) and amendment of the brine to limit the accumulation of sulfate,
  - Disposal as discussed above,
  - Reintroduce the discharged brine to the treated water. This would result in a final sulfate concentration equal to that of the untreated water, and a slight increase in TDS as a result of the additional NaCl addition,
- Treatment the brine to remove sulfate,
  - Treatment by nanofiltration membranes would result in a large portion of the brine being ready for reuse, and a smaller fraction loaded with sulfate that must then be discharge above.

Counter-current regeneration was carried out at a flow rate of 0.37 gpm for 6.4 BV (salt loading rate of 25 lbs/ft^3) using a 6 percent NaCl solution. Spent brine from the regeneration was treated with the PNDM to remove perchlorate and nitrate. Complete removal of both perchlorate and nitrate to below analytical detection limits was observed for all 30 runs. Sulfate and bicarbonate, however, were not removed by the PNDM. As a result, their respective concentrations in the brine increased with each consecutive regeneration. Figure 6.3 displays the major anions present in the composite brine following a regeneration procedure. Perchlorate and nitrate concentrations remained consistent for all 30 runs averaging 3150 µg/L and 280 mg/L, respectively. The slight variability in the increasing sulfate and bicarbonate trend is associated with the variable influent concentrations of respective anions. During Run 15, extra rinse water (normally directed to waste) was inadvertently combined with the spent brine, thereby diluting the concentration of anions. As a result of this incident, both sulfate and bicarbonate were both reduced by nearly 2000 mg/L before continuing to increase with each consecutive run. The concentrations of bicarbonate and sulfate in the composite brine appeared to be approaching a steady state concentration during the last 10 cycles.

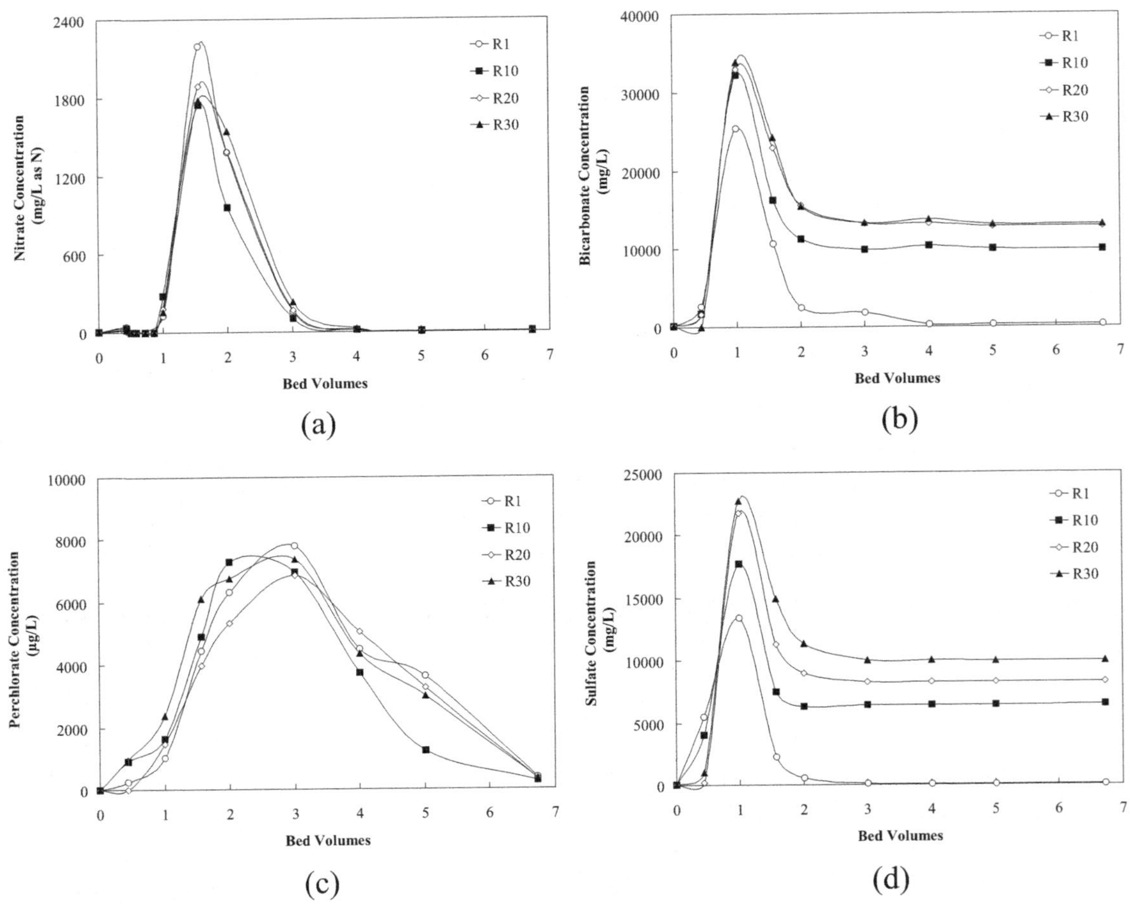

Figure 6.4 Detailed regeneration profiles during chemical brine treatment (polyacrylic resin)

Detailed regeneration profiles of nitrate and perchlorate for selected runs (Runs 1, 10, 20, and 30), as seen in Figure 6.4 (a) and Figure 6.4 (c), were nearly identical to the results found during baseline testing. Complete removal of perchlorate and nitrate in the spent brine ensured that the treated brine could successfully be reused numerous times without a significant loss in resin performance. In addition, the accumulation of sulfate and bicarbonate in the treated brine had no effect on the regeneration performance of perchlorate and nitrate when recycled. In the following Figure 6.4 (b) and Figure 6.4 (d), the effect of the accumulation of bicarbonate and sulfate in the spent brine and on the resin can be seen in regeneration profiles for selected runs (Run 1, 10, 20, and 30).

WATER QUALITY

Three major sampling campaigns were performed to characterize the water quality of the ion-exchange system utilizing chemical brine treatment and recycling. Particular attention was given to the brine water quality before (spent brine) and after chemical brine treatment (treated brine).

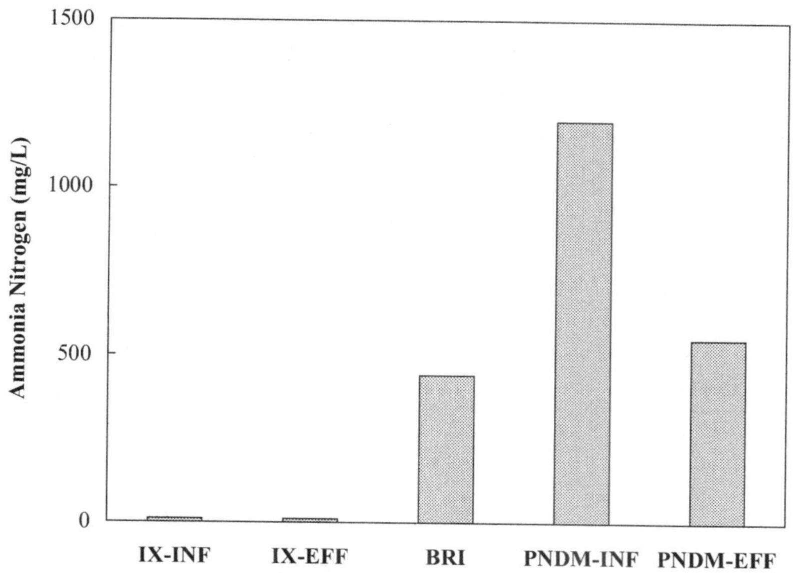

Figure 6.5 Profile of ammonia (redundant) through the treatment process

Ammonia was amended to the spent brine as a reductant before being processed by the PNDM. To ensure that reductant was available in excess for the PNDM and to verify that the excess reductant would not affect the treated water quality during exhaustion of the resin, ammonia was monitored throughout the treatment process. As shown in Figure 6.5, the PNDM was amended with approximately 1200 mg/L NH_4^+-N, corresponding to the stoichiometric requirement based on perchlorate and nitrate in the spent brine plus a targeted 500 mg/L residual. It is important to note that no ammonia was detected in the treated effluent from the ion-exchange column at any time during exhaustion, even though the resin was regenerated with recycled brine containing 500-600 mg/L NH_4^+-N. These data indicate that 3 BV of rinse water was sufficient to remove any residual ammonia.

Table 6.1 summarizes selected water quality parameters that had a noticeable increase over the course to the 30 cycle treatment and recycle experiment. Sulfate and bicarbonate concentration, as expected, accumulated in the treated brine as the PNDM only chemically removed perchlorate and nitrate. Other parameters included uranium, barium, fluoride, and silica. These elements, were measured in low to non-detect amounts in the raw water, accumulated after each regeneration and were not removed by the PNDM. Additional analyses performed on both the recycled brine and the treated ion exchange effluent indicated non-detect levels of the catalyst.

Total arsenic was measured to be 1.5 μg/L in the raw water. As arsenic was adsorbed by the resin during exhaustion and displaced during regeneration, the concentration of arsenic measured in the spent brine was 60-70 μg/L (Figure 6.6). After the spent brine was treated with the PNDM, however, arsenic was measured below detection limits (<1 μg/L) in all samples. It was hypothesized that the arsenic was co-precipitating out within the PNDM.

To investigate the potential co-precipitation of arsenic, the total hardness concentrations were analyzed. As shown in Figure 6.7, the total hardness measured in the spent regenerated brine averaged 100 – 175 mg/L as $CaCO_3$, despite using low hardness salt (Morton Culinox 999) for preparation of the initial sweet brine and subsequent salt amendments of treated brine. After brine treatment with the PNDM, hardness was not detected in any samples. After visual inspection of

Table 6.1
Chemically treated brine water quality parameters with noticeable increase

Parameter	Run 1	Run 30
Sulfate	1600 mg/L	9000 mg/L
Bicarbonate	4000 mg/L	16600 mg/L
Uranium	175 µg/L	1700 µg/L
Barium	90 µg/L	120 µg/L
Fluoride	1.5 mg/L	7.7 mg/L
Silica	28 mg/L	44 mg/L

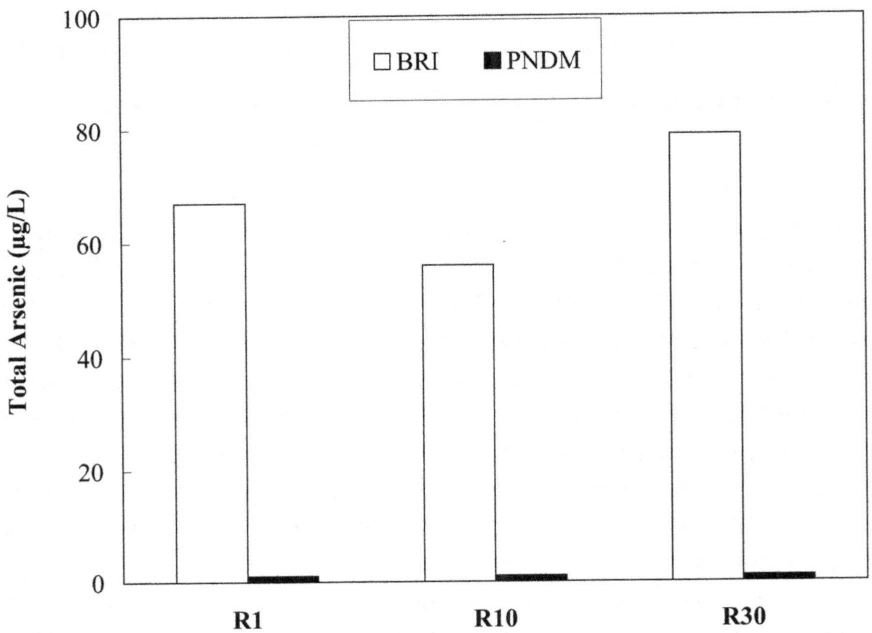

Figure 6.6 Arsenic removal during chemical brine treatment

the PNDM lines during system maintenance events, precipitation was confirmed by the noticeable scale buildup. Since significant scaling was observed, it was hypothesized that arsenic was removed through adsorption and occlusion with calcium carbonate and magnesium hydroxide precipitates. Depending on the water quality, the co-precipitation of arsenic or other contaminants with $CaCO_3$ could be a serious concern.

OPERATION AND MAINTENANCE

The PNDM was successful at completely reducing perchlorate and nitrate in the spent ion-exchange brine. Spent ion-exchange brine was corrosive in nature with a high salt content (6 percent NaCl) and a pH of 9. Consequently, Hastelloy was the material of construction for the

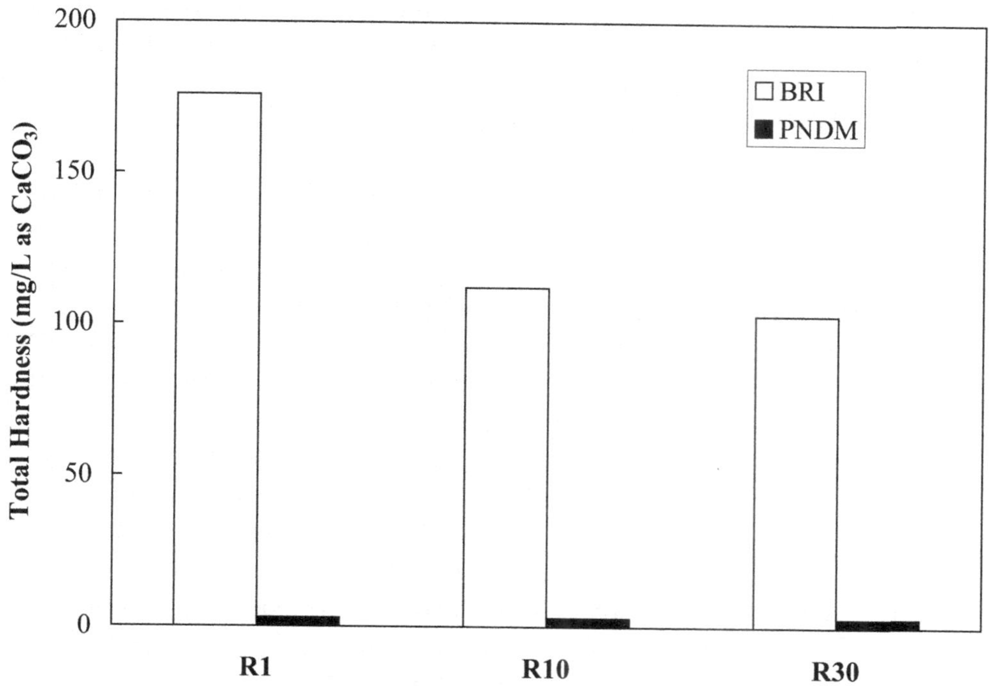

Figure 6.7 Hardness levels during chemical brine treatment

PNDM for its excellent corrosive and heat resistant properties. Due to the precipitation of $CaCO_3$, however, complete clogging of the system was occasionally encountered while processing spent ion-exchange brine. Maintenance to correct problems associated with scaling (and resultant clogging) was the primary impediment to continuous operation.

Cleanup of the PNDM would appear to be quite simple as an acid solution would be required to effectively remove precipitation in the lines. Since a blockage in the flow was the primary system failure, no clean water could be pumped through the system to cool PNDM from 230°C prior to disassembly. As a result, approximately four hours were required to allow the system to cool to a safe temperature and depressurize from 600 psi for disassembly. Once the system temperature had reached ambient and the system was depressurized, each leg of the tubing was revealed by removing heat insulation, the appropriate fittings were removed, and the lines causing the blockage in flow were flushed with acid. Next the system was reassembled and carefully leak tested to ensure that the system would maintain pressures up to 600 psi before resuming operation. The entire system could not be periodically cleaned with an acid solution because of the catalyst-coated media contained within the PNDM reactor. Each leg of tubing leading up to and after the reactor needed to be removed to identify the blockage. Blockage removal turned out to be a time-consuming task.

PNDM CONCLUSIONS

Calgon's PNDM system was found to quickly and effectively destroy perchlorate and nitrate in a concentrated spent ion-exchange brine, to below analytical detection limits. Additionally, both resin and PNDM performance were consistently maintained when treated brine was recycled 30 times. Sulfate and bicarbonate accumulating in the brine to 9,000 mg/L and 16,600 mg/L, respectively, did not affect the effluent water quality after each recycle.

Successful operation of the PNDM, however, required careful monitoring to identify and prevent the following water quality and O&M issues:

- Although, sulfate and bicarbonate accumulated in the treated and recycled brine without deleterious effects, uranium accumulated in the recycled brine from 175 μg/L to 1700 μg/L. The accumulation of heavy metals, radiologicals, or other constituents depend on water quality and may pose disposal and reuse issues.
- A build-up of scale was observed in the heat exchanger following the PNDM reactor even when low hardness salt was used. This can lead to process failures due to the clogging of the system.
- Since scale build-up is associated with carbonate removed by ion exchange, a potential solution is to optimize the ion exchange process (i.e., process configuration, acid rinse of the resin before regeneration, etc.) such that carbonate removed from the water is minimized and therefore less would be present in the spent brine.
- Due to the corrosive nature of brine and PNDM operating conditions, robust and compatible materials of construction must be appropriately utilized.
- The pilot-scale PNDM operated under elevated temperature and pressure conditions (230°C and 600 psi), and was found to require lengthy start-up and shut-down times.

CHAPTER 7
ELECTROLYTIC BRINE TREATMENT

The IXL™ process was developed by Ionex Ltd. (Gloucestershire, UK) for the electrochemical treatment of concentrated ion exchange brines. Although the electrolytic cell was developed for the removal of nitrate, it was anticipated that perchlorate reduction was potentially possible. As a result, laboratory trials were performed to evaluate the capability of the electrolytic process to reduce perchlorate. Additionally, pilot-scale testing was performed to validate bench-scale experiments.

The IXL™ cell employed two distinct processes for the destruction of these nitrate and/or perchlorate contaminants: the 'Activation' or 'Sweep cycle' and the 'Sweep-Hold cycle.' Both processes were run independently of each other. The 'activation' cycle involved the cell voltage sweeping continually within controlled voltage parameters at a controlled sweep rate. This reconstructs the surface of the electrode to preferentially destroy nitrate, increases electrical efficiency and reduces by-product production to a minimum.

The sweep-hold cycle was entered after the electrodes had been activated, and increased the cell's electrical efficiency and nitrate destruction rate. The voltage was 'held' at the voltage where efficient nitrate destruction occurred, as opposed to the sweep cycle, which only entered these zones for a brief period during the voltage cycle. In order to maintain the highly active electrode surface, the voltage was periodically swept through a small number of 'activation' cycles. The function of this was to prevent any potential surface poisoning effects and maintain the electrode surface in optimum condition.

ELECTROLYTIC BENCH-SCALE INVESTIGATION

Bench-scale testing was conducted using a laboratory test cell comprised of a working electrode of titanium wire coated with a 1 μm thickness of rhodium as an industrial finish electroplated layer. The nominal surface area of this electrode was 2 cm^2. The electrode had been previously 'activated' (an Ionex proprietary process which increased the surface area of the electrode to approximately 240 times the planar area and also templates the surface at an atomic scale to favor nitrate reduction). This electrode was contained within a polypropylene test cell containing 250 cm^3 of electrolyte maintained at 60°C in a water bath. The counter electrode was a titanium sheet coated with ruthenium dioxide and titanium dioxide (DSA) of approximately 20 cm^2 area. The cell also included a solid polymer reference electrode with a reference voltage similar to a standard calomel electrode. The voltage between the cathode surface and the reference electrode was used to control the applied potential to the cell. Cell voltages and current were logged to a PC via a Microlink data acquisition system running Microlink Windmill software.

Impact of Perchlorate on Nitrate Reduction

Preliminary experiments were conducted to determine if there were any deleterious effects on the nitrate reduction process in the presence of perchlorate. Results showed that no degradation of the electrode surfaces occurred and no reduction in the nitrate destruction rate was observed with perchlorate concentrations as high as 570 mg/L. From previous detailed reaction kinetic studies in the alkaline potassium chloride brine, it had established that the nitrate ion adsorption conforms to quasi-Langmuirian kinetics. At a bulk concentration in excess of approximately 4,000 mg/L, the electrode surface was essentially saturated with nitrate ions. Consequently, significant perchlorate reduction would not be expected at higher nitrate concentrations.

Dependence of Nitrate on Perchlorate Reduction

In a second series of experiments, several runs were performed with different initial levels of nitrate in the electrolyte (2 M KCL + 0.2 M KOH) to determine if perchlorate reduction would occur when nitrate levels were below 4,000 mg/L. Results of three specific test runs are summarized in Table 7.1. In the first run, an initial nitrate level of approximately 18,000 mg/L was reduced to 10,000 mg/L in the presence of 50 mg/L perchlorate. As expected, little perchlorate reduction (<1 percent) was observed in the presence of 10,000 mg/L nitrate. A second run was performed with similar initial conditions, but nitrate levels were reduced to 5,000 mg/L, and up to 6 percent perchlorate reduction was observed. In the final run, nitrate levels were reduced to 2,000 mg/L and up to 16 percent perchlorate reduction was observed.

These tests confirmed that little perchlorate reduction would be observed until the nitrate concentrations were reduced to below 4,000 mg/L. It is important to note that, in these tests, the electrodes were continually cycled in 'activation mode.' Therefore, a final test was performed to evaluate the adsorption kinetics of perchlorate on the rhodium surface by conducting cyclic voltametry measurements at varying sweep rates in the base electrolyte containing approximately 6,000 mg/L perchlorate with no nitrate present. At slower sweep rates, a clear perchlorate reduction peak was observed, which confirmed the reported slow adsorption kinetics. As shown in Figure 7.1, the peak occurred at approximately -950 mV, at a similar potential to the nitrate reduction peak. Under normal cell operation, in 'sweep-hold' mode, it would be expected that significant reduction of perchlorate could occur at low nitrate levels.

ELECTROLYTIC PILOT-SCALE EVALUATION

A demonstration plant, complete with an ion-exchange system and pilot-scale electrolytic brine treatment cell, was built and shipped to the pilot testing site for evaluation. The ion exchange system contained three pilot-scale ion-exchange columns, each containing 20 liters of resin which could be operated in a series configuration to remove nitrate from a raw water source. A blending control of the water through the ion exchange plant would allow the level of nitrate or perchlorate to be controlled to a desired output concentration. Attention was only given to the operation of the electrolytic cell for this study.

The pilot-scale electrolytic cell was a bipolar design and contained two end electrodes with seven bipolar electrodes of dimensions 220 by 300 mm. Six solid polymer reference electrodes allowed the operator to monitor the voltages of the working electrodes. Laboratory studies revealed that in order to fully activate the electrodes, the cell sweep rate and voltage limits must be tightly controlled. The demonstration plant used a switched mode power supply (PSU) with reference electrodes fed into the PLC and an analog output taken from the PLC to the PSU. The sampling rate of the PLC, however, was insufficient to accurately control the cell parameters, which meant that the electrodes in the cell were only partially activated.

Table 7.1
Laboratory perchlorate reduction data

Sample	Comments	Nitrate conc. (ppm)	Perchlorate conc. (ppm)	Average (ppm)	% ClO_4^{2-} reduced (%)
641	PMD-Z53 electrode; Z67; solution ex-	14632	44.56	44.69	0.96
643	Z61 ~13,500 ppm; duplicate samples	14854	44.82		
642	Z67; solution ex-Z60 ~8,900 ppm	9874	43.65	44.26	
644	as 642	10129	44.86		
645	Stock matrix; ~41 ppm ClO_4^{2-}	18917	46.53	Control samples	
646	Spike – as 645 + 10 ppm ClO_4^{2-}	19086	56.47		
651		15530	40.89	40.67	6.22
652		15464	40.45		
653		4971	38.61	38.14	
654		4999	37.66		
721		2100	37.9	36.85	16.15
722		2100	35.8		
723		6000*	44.3	43.95	
724		6000*	43.6		

* Ionex data.

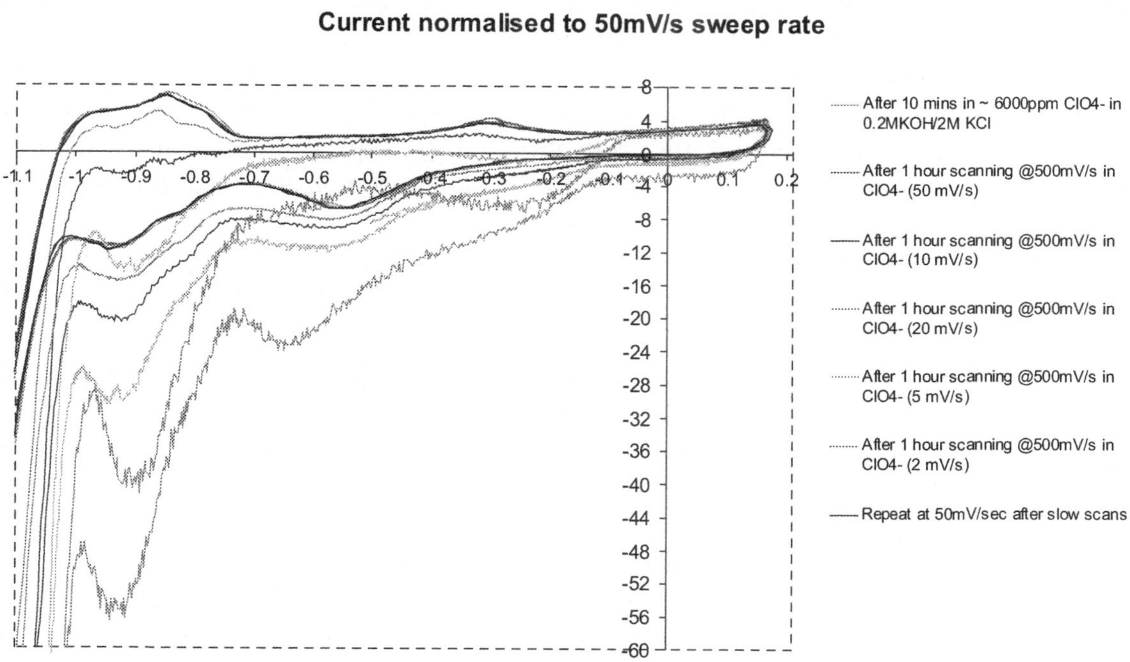

Figure 7.1 Perchlorate reduction at different scan rates (scan rate bracketed in legend)

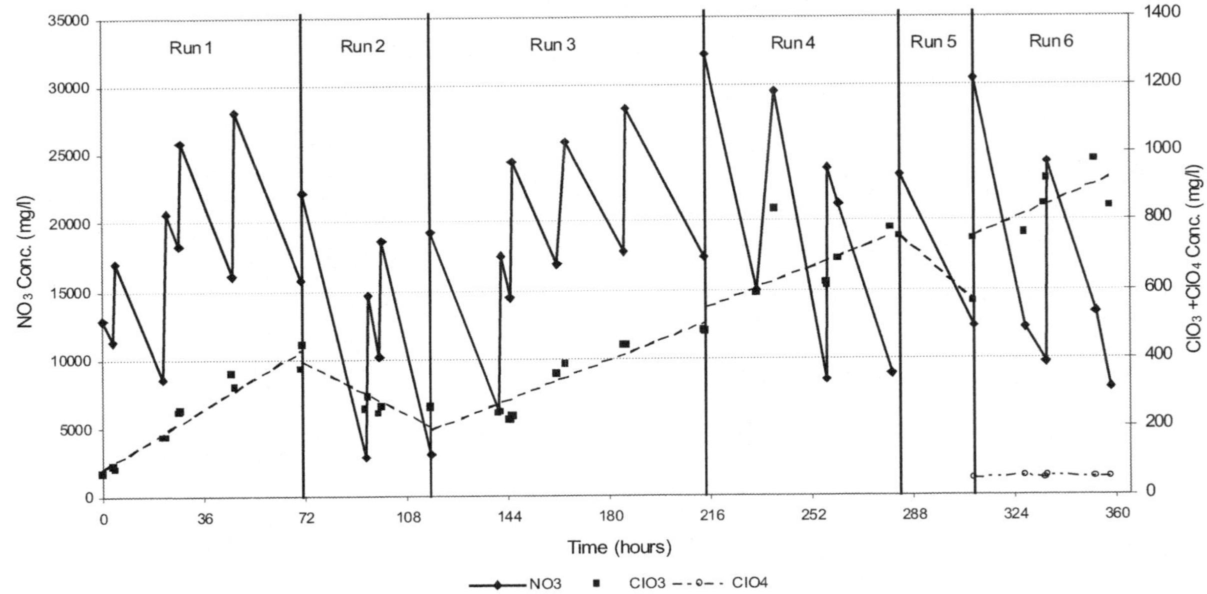

Figure 7.2 Overview of cell conditioning and activation

Initial Cell Conditioning and Activation

The electrodes in the cell had a one-month partial 'activation' prior to being shipped. In order to continue the electrode 'activation,' nitric acid additions were made to the base solution of 2 M KCl and 0.2 M NaOH to maintain the nitrate levels between 18,000 and 8,000 mg/L.

Nitrate destruction was observed under both 'activation' and 'sweep-hold' operation (Figure 7.2). As predicted from the laboratory data, operation in 'sweep-hold' reduced nitrate at approximately twice the rate compared to operation in 'activation' mode only and also gave superior electrical efficiencies. As this test series progressed, the electrical efficiency of the cell continued to increase, indicating that the electrodes were continuing to 'activate.' During 'sweep-hold' operation, the rhodium surfaces were held at a mean voltage of -925 mV for three minutes, followed by five voltage cycles through the 'activation' range. Typical power consumption ranged between 10 to 20 Whr/g NO_3^-. The laboratory studies demonstrated nitrate reduction at <3.5 Whr/g was achieved at slightly lower hold voltages.

As the development process was ongoing, the cell used on the trial had not been optimized for perchlorate removal. Three limitations were identified to affect pilot-scale cell performance:

- The cell voltage was operated and controlled by the programmable logic controller (PLC), which automated the plant operation. The sampling rate of the PLC limited the accuracy of control over the 'activation' parameters. Previous laboratory data had shown that these parameters needed to be precisely controlled to achieve full 'activation' of the electrodes. This was addressed by the incorporation of a high-speed dedicated instrumentation and control board into the PSU.
- The demonstration cell had only seven bipolar electrodes. This meant that significant 'end electrode effects' occurred. This resulted in an uneven voltage profile across the

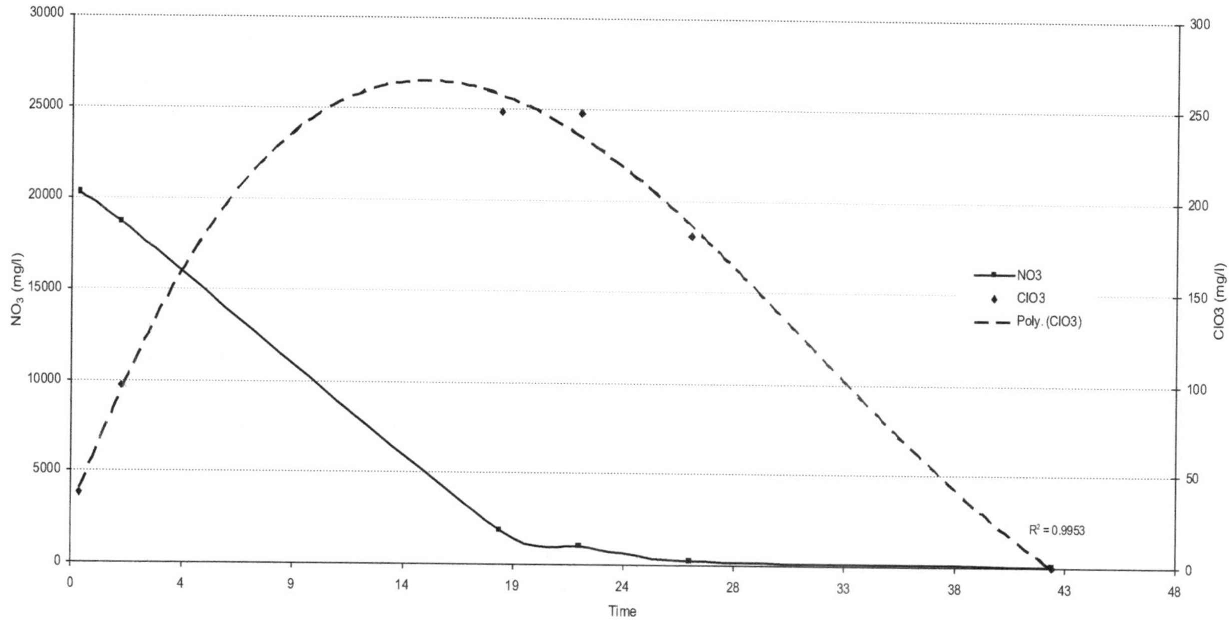

Figure 7.3 Chlorate reduction at low nitrate concentrations

electrode pairs in the cell. As full-scale cells would utilize close to 100 electrodes, end electrode effects are expected to be insignificant.

- The pilot-scale electrodes in the cell were smaller than those expected to be used in the commercial cells. This meant that electrode edge effects were proportionally higher and may have been responsible for inefficiencies in cell operation.

With the exception of runs 2 (sweep-hold) and 5 (activation), chlorate production was observed, increasing to a final concentration of 840 mg/L. As the sweep and hold parameters were varied during these tests, the formation of chlorate was considered more likely to be due to over potential on the anode surfaces as a result of inadequate control through the PLC, rather than an inherent problem in the cell. This was supported by the observation that, in runs where the process parameters were maintained closer to the desired values, chlorate reduction was observed. Continued investigations are being performed under laboratory conditions.

In Run 6, 50 mg/L perchlorate was added to the bulk solution. However, the nitrate concentration was not reduced to below 4,000 mg/L and, as predicted from laboratory data, no perchlorate reduction was observed.

Evaluation of (Per)chlorate Reduction at Low Nitrate Conditions

Once the cell had been sufficiently activated, a fresh solution was used to determine if chlorate build-up during the early stages of a process cycle could be reversed by processing to low nitrate levels. As shown in Figure 7.3, the initial nitrate concentration was 20,000 mg/L. After 18 hours, this had been reduced to below 2,000 mg/L. At the lower nitrate levels, the electrical efficiency of the cell deteriorated and chlorate levels were reduced to zero, but less than 10 percent perchlorate reduction was observed (Figure 7.4). It is likely that the chlorate would

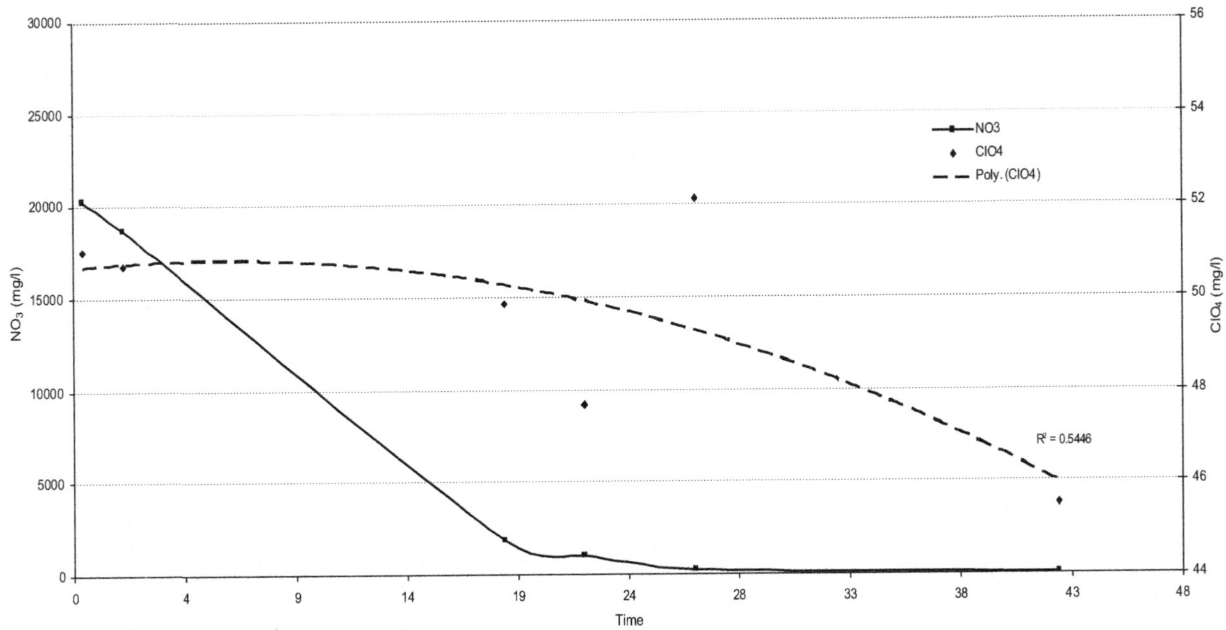

Figure 7.4 Perchlorate reduction at low nitrate concentrations

be preferentially reduced to perchlorate and this was considered likely to explain the poor perchlorate reduction. The electrical efficiency of nitrate reduction was 12 Whr/g at concentrations greater 1,000 mg/L. The overall destruction rate was 30 g/h.

ELECTROLYTIC REDUCTION CONCLUSIONS

The goal of the electrochemical testing was to determine whether or an electrochemical cell could reduce perchlorate and nitrate in spent ion exchange brine. Initial bench-scale investigations confirmed that nitrate reduction was not inhibited in the presence of perchlorate. It was also determined that perchlorate reduction could occur. However this was only observed if nitrate levels were below 2,000 mg/L.

Proceeding with pilot-scale testing, the cell was initially operated to activate and condition new electrodes. Interestingly, chlorate production was observed in the presence of high nitrate (4,000 to 20,000 mg/L). Additional tests demonstrated that chlorate production could be reduced in the presence of low nitrate (less than 2,000 mg/L), though perchlorate removal was still limited to less than 10 percent.

The electrolytic process was demonstrated to be an effective means to remove nitrate from spent ion exchange brine. However, reasonable perchlorate reduction rates were not achieved. Although initial testing (bench- and pilot-scale) of the electrolytic cell for perchlorate removal showed promise, further development and optimization of this technology is required before it can be applied effectively to perchlorate-contaminated water.

CHAPTER 8
ENGINEERING ANALYSIS

Extensive bench- and pilot-scale testing have clearly demonstrated that spent brine from a conventional ion exchange process used to remove perchlorate from contaminated water can be treated and reused. The objective of this chapter is to evaluate the cost of brine treatment and reuse. The costs presented in this chapter are for a complete full-scale treatment system including the ion exchange system, necessary equipment for applicable brine management options, construction-related services, and annual O&M. It should be noted that the focus of the engineering analysis is not the ion exchange process, but the relative associated costs to manage the brine after the regeneration cycle.

COST MODEL

A cost model was developed to determine the capital, operation, and maintenance costs associated with removing perchlorate from groundwater using a non-proprietary ion exchange system along with one of multiple forms of brine management. The brine management options explored in this chapter include brine treatment through the 3 or 6 percent brine biological treatment, Calgon Carbon Corporation's (Calgon) Perchlorate Nitrate Destruction Module (PNDM), brine line disposal (no brine treatment), or a combination of brine treatment with treated brine disposal through a dedicated brine line. These scenarios were evaluated as three brine management alternatives discussed in this section.

Alternative 1 – No Brine Treatment

In Alternative 1, the brine from a conventional ion exchange system is used once to regenerate the spent resin from the ion exchange unit. The waste brine is then sent directly to a dedicated brine line for disposal. In this scenario, the brine is not treated or recycled prior to disposal. Therefore, the waste brine discharged contains concentrated levels of both perchlorate and nitrate.

Alternative 2 – Brine Treatment and Reuse

In Alternative 2, the spent brine from a conventional ion exchange system is treated to destroy the perchlorate and nitrate. The treated brine is then sent to the sweet brine tank and a small quantity of salt is added to restore the brine strength. Overall, this scenario results in a substantial reduction in the quantity of waste brine discharged while dramatically improving the quality of the discharged brine (perchlorate and nitrate removed).

For brine treatment, both the PNDM and biological methods are evaluated. To date, the challenge with biological brine treatment has been the acclimation of a biological culture to reduce perchlorate in a high salt concentration. For this reason, both a 3 and 6 percent brine solution have been evaluated in this alternative. In total, Alternative 2 consists of the conventional ion exchange system with three possible brine treatment options. The brine treatment options are:

1. 3 percent brine biological treatment,
2. 6 percent brine biological treatment,
3. 6 percent brine PNDM treatment.

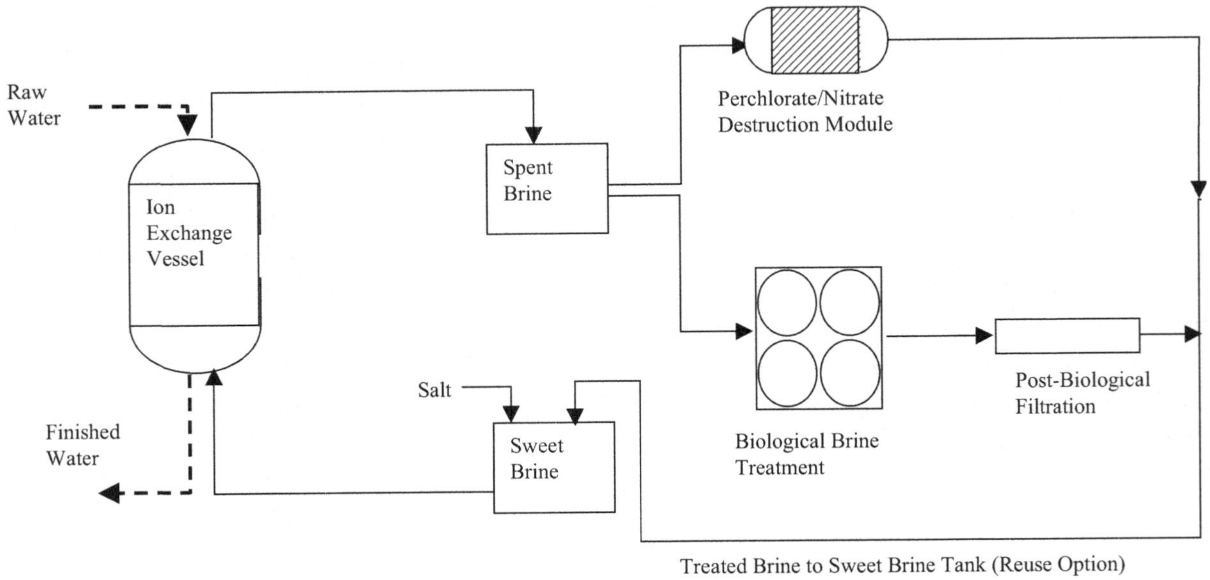

Figure 8.1 Schematic of ion exchange with brine treatment and reuse options

Figure 8.1 presents the ion exchange system and the three brine management options of Alternative 2. In the schematic, the effluent from the brine treatment is recycled to the sweet brine tank.

Alternative 3 – Brine Treatment and Disposal

The third alternative explores the option of treating the brine and then disposing the treated brine through a brine line without recycling. This alternative assumes the discharge of perchlorate-laden brine would not be allowed and the appropriate permitting for recycling treated brine could not be obtained. Both biological and PNDM treatment were evaluated for 6 percent brine in this alternative.

Summary of Alternatives

Table 8.1 provides a summary of the estimated capital and O&M costs for the three treatment alternatives explored using the cost model. Input variables for the cost model were derived from field results, theoretical considerations for each brine treatment design, and estimates provided by equipment manufacturers. The total water production cost ($/1000 gal and $/acre-foot) for each alternative are graphically presented in Figure 8.2.

Of all the alternatives evaluated, the 6 percent biological brine treatment with recycle was the least expensive treatment scenario. Water production costs associated with the 6 percent brine biological treatment and brine reuse was $1.00/1000 gal ($326/AF). The 3 percent system was just slightly higher at $1.19/1000 gal ($387/AF). The 3 percent brine biological treatment is slightly more expensive due to the increased volume and number of biological reactors and additional pumping requirement for processing the 3 percent brine solution.

Table 8.1
Life cycle cost for conventional ion exchange with brine management alternatives

Brine management alternative	Description	O&M cost ($/AF)	Capital costs* ($/AF)	Total cost ($/AF)	Total cost ($/1000gal)
1	No Treatment & Brine Line Discharge	317	85	402	1.23
2	6% Brine – Biological Treatment & Reuse	220	106	326	1.00
	3% Brine – Biological Treatment & Reuse	259	127	386	1.19
	PNDM Treatment & Reuse	291	172	463	1.42
3	6% Biological Treatment & Brine Line Discharge	346	107	453	1.39
	3% Biological Treatment & Brine Line Discharge	386	129	515	1.58
	PNDM Treatment & Brine Line Discharge	419	173	591	1.82

* Amortized for 20 years at 5% interest.

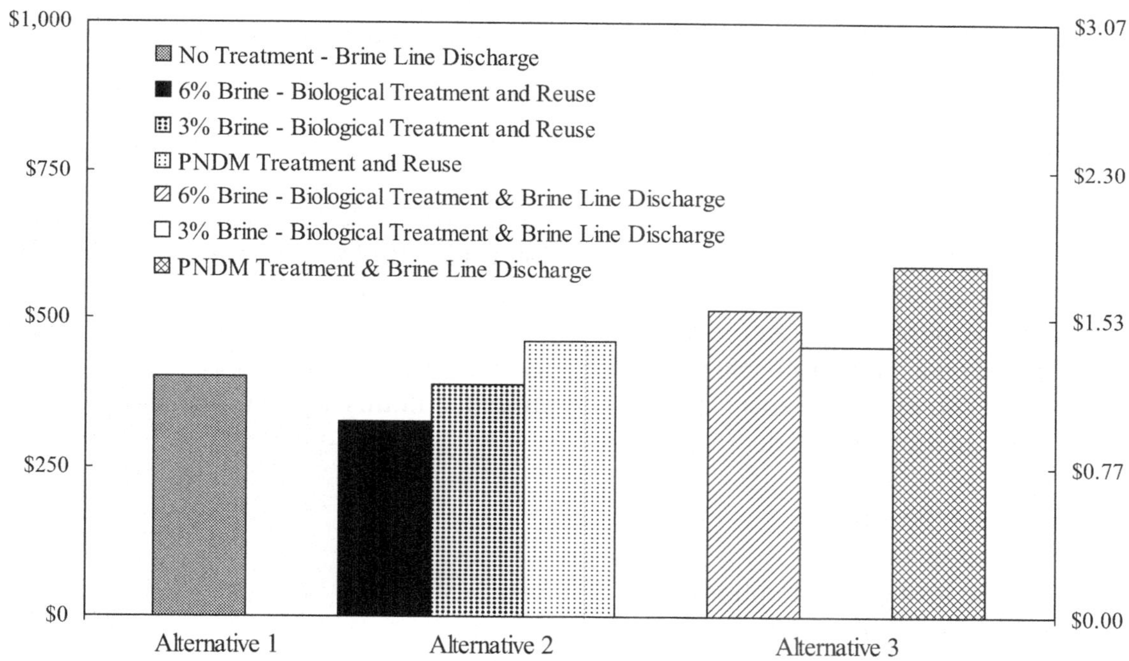

Figure 8.2 Life cycle cost for 2500 gpm conventional ion exchange treatment plant with brine management alternatives

By recycling the treated brine (Alternative 2) the anticipated annual salt costs were reduced from $600,000/year to $90,000/year for both the biological and PNDM brine treatment options. Some or all of these cost savings were offset by an increase in the amortized capital costs of approximately $80,000/yr for 6 percent brine biological and $350,000/year for PNDM. The capital costs associated with the PNDM brine treatment increased the life cycle from $1.23/1000 gallons ($402/AF) for the brine line discharge option (no brine treatment) to $1.42/1000 gallons ($463/AF) for the PNDM brine treatment with the brine recycle option.

The third alternative (brine treatment and disposal without reuse) generated high treatment costs due to added expense of disposing the brine after treatment, for both the PNDM and biological treatment options. The 6 percent brine biological treatment with brine line disposal option was $1.39/1000 gallons ($453/AF), while the PNDM option was $1.82/1000 gallons ($591/AF).

There are intangible benefits to Alternative 2 and 3 not realized in Alternative 1. Directly disposing of the spent brine to a dedicated brine line does not destroy the perchlorate. The perchlorate is simply moved from one aqueous environment to another. Future regulations may require the brine to be treated prior to disposal and would therefore require a plant upgrade to meet the future regulations. For a similar life-cycle cost, Alternative 2 is cost competitive with Alternative 1. An advantage of Alternative 2 is that it ensures that the plant is taking action to prepare for possible future regulations and litigation, and is environmentally conscious with regard to brine management. While not incorporated in these alternatives, it is anticipated that brine discharge costs will increase over time. As waste disposal costs increase, waste reduction becomes increasingly more attractive.

COST MODEL CALCULATIONS

Full-scale treatment-cost projections used in the cost model were calculated from a combination of standard engineering practice and the pilot results. Biological brine treatment costs were estimated from assumptions used to develop a conceptual full-scale biological brine treatment system based on pilot design and results. PNDM treatment costs were provided by Calgon and estimated based on existing full-scale design experience. The assumptions used in the engineering analysis for the ion exchange cost estimates are presented in this section. The major components of each cost evaluation included:

- Design flow rates
- Water quality
- Brine management options
- Regeneration requirements
- Capital costs
- Ancillary treatment costs
- Operation and maintenance (O&M) costs
- Labor costs
- Construction related costs

Capital Cost

Capital costs are those items that contribute to a new ion exchange facility, including treatment equipment, installation, and other construction-related costs. The estimated capital costs for

Table 8.2
2500 gpm estimated capital and construction related costs

Capital cost parameter	No brine treatment*	6 percent biological brine treatment†	3 percent biological brine treatment†	PNDM brine treatment‡
Brine Line Connection	$40,000	—	—	—
Conventional IX	$2,000,000	$2,000,000	$2,000,000	$2,000,000
IX Installation	$400,000	$400,000	$400,000	$400,000
PNDM	—	—	—	$2,000,000
PNDM Installation	—	—	—	$700,000
Biological	—	$330,000	$650,000	—
Biological Installation	—	$162,000	$325,000	—
Facility Structure	$240,000	$430,000	$620,000	$290,000
Construction Related Costs	$965,000	$1,200,000	$1,400,000	$1,900,000
EL&A Costs§	$292,000	$361,000	$435,000	$586,000
Total Cost to Construct	$4,300,000	$5,300,000	$6,400,000	$8,600,000
Amortized Capital Cost	$345,000/yr	$428,000/yr	$515,000/yr	$695,000/yr
Amortized Cost per Acre-Foot**	$85/AF	$106/AF	$127/AF	$172/AF
Cost per 1000 Gallons	$0.26	$0.32	$0.39	$0.53

* Alternative 1 – No brine treatment with brine line discharge.
† Alternative 2 – Biological brine treatment with brine reuse.
‡ Alternative 3 – PNDM brine treatment with brine reuse.
§ Engineering, Legal and Administrative costs.
** 2,500 gpm system at 5% interest for 20-years.

a 2500 gpm perchlorate removal system are presented in Table 8.2. Capital costs are presented for the ion exchange system, with and without brine treatment. The major components of the capital costs associated with a new ion exchange plant and the estimated percentage of the total cost to construct the new plant are presented in Table 8.3.

The breakdown of the capital cost is relatively similar for each treatment option. In each case, the equipment contributes the largest percent of the capital costs. The three remaining categories are directly related to the construction of the new facility. The assumptions used for the general construction-related services are presented in Table 8.4 and represent a percentage of the capital equipment costs.

Operation & Maintenance Costs

Annual O&M costs are associated with daily consumables. These consumables include utilities and daily maintenance requirements. For the cost analysis all equipment life assumed to be 20 years with resin life estimated at 5 years. This data is presented in Table 8.5 and represents the costs associated with a production rate of 2500 gpm. The O&M costs do not include associated

Table 8.3
Capital cost components

Cost component	No brine treatment*	Biological brine treatment with brine reuse†	PNDM brine treatment with brine reuse‡
Treatment Related Equipment	62%	63%	62%
General Construction Related Costs	22%	22%	23%
Engineering, Legal and Administrative	7%	7%	7%
Contingency	9%	8%	8%

* Alternative 1 – No brine treatment with brine line discharge.
† Alternative 2 – Biological brine treatment with brine reuse.
‡ Alternative 2 – PNDM brine treatment with brine reuse.

Table 8.4
Assumptions for constructed related services

Construction related services	Percentage of capital costs
Electrical =	8%
Piping =	5%
Instrumentation =	8%
Civil Site Work =	5%
Contractor OH&P, Bonds, and Insurance =	10%
EL&A* =	8%
Construction Contingency =	10%
Interest Rate =	5%

* EL&A = Engineering, legal and administrative services.

costs for water quality testing or other engineering related costs due to systems unrelated to the ion exchange system. The major components of the O&M costs are presented in Table 8.6.

For the brine line with no brine treatment option, the salt cost was the largest percent of the annual O&M costs. Due to the recycling capabilities associated with the biological and PNDM brine treatment options, the salt consumption rates were greatly reduced and, therefore, comprised a smaller percentage of the total O&M costs. The utility cost was a large component for each brine management option, ranging from 22 to 33 percent of the total O&M cost. Both the biological and PNDM had high electricity costs associated with the technologies. However, these technologies had lower O&M costs in other areas (i.e. salt and other consumables) which helped to reduce the overall costs.

Table 8.5
2500 gpm estimated annual O&M costs

O&M parameter	No brine treatment*	6 percent biological brine treatment†	3 percent biological brine treatment†	PNDM brine treatment‡
Salt	$600,000	$90,000	$90,000	$90,000
Maintenance	$270,000	$340,000	$450,000	$545,000
Seeding, Nutrients, Chemicals	—	$46,000	$92,000	$50,000
Resin Costs	$30,000	$30,000	$30,000	$30,000
Power	$285,000	$290,000	$295,000	$375,000
Labor (+overhead)	$70,000	$92,000	$92,000	$92,000
Brine Disposal	$25,000	—	—	—
Total Cost per Year	$1,300,000	$888,000	$1,050,000	$1,177,000
Cost per Acre-Foot	$317/AF	$220/AF	$259/AF	$291/AF
Cost per 1000 Gallons	$0.97	$0.67	$0.80	$0.89

* Alternative 1 – No brine treatment with brine line discharge.
† Alternative 2 – Biological brine treatment with brine reuse.
‡ Alternative 3 – PNDM brine treatment with brine reuse.

Table 8.6
O&M cost components

Cost component	No treatment with brine line*	Biological brine treatment with brine reuse†		PNDM brine treatment with brine reuse†
		(3%)	(6%)	
Salt	47%	10%	9%	7%
Utilities	22%	33%	29%	30%
Miscellaneous O&M‡	21%	39%	41%	44%
Labor	6%	11%	9%	12%
Resin Replacement	2%	3%	3%	2%
Brine Line Disposal	2%	< 1%	< 1%	< 1%
Other§	0%	3%	8%	4%

* Alternative 1 – No brine treatment with brine line discharge.
† Alternative 2 – Brine treatment with brine recycle.
‡ Miscellaneous O&M covers additional items not specifically listed, including general equipment and facility maintenance.
§ "Other" refers to technology specific items. Biological brine treatment includes seeding and other nutrients, membrane cleaning, etc. For the PNDM process is refers to chemicals specific to the PNDM process.

Table 8.7
Summary of engineering assumptions used in the ion exchange cost analysis

Input Design Information	*Value*
Design Flow Rate =	2500 gpm
Design Run Time* =	400 BV
Service Loading Rate =	4 gpm/cu-ft
Salt Loading Rate =	25 lbs/cu-ft
Salt Strength =	6%
Number of Redundant Ion Exchange Vessels =	1 Vessel
Capital Costs Associated With the Ion Exchange System	***Value***
Cost of 8.5-ft diameter Vessel =	$168,000
Cost of 10-ft diameter Vessel =	$231,000
Facility Structure =	$150/sq-ft
Auxiliary Equipment =	$800,000
O&M Costs Associated With the Ion Exchange System	***Value***
Plant Equipment Life =	20 years
Resin Unit Cost =	$175/cu-ft
Resin Life =	5 yrs
Salt Unit Cost =	$110/ton
Energy Unit Cost =	$0.12/kW-hr
Pumping Energy Efficiency =	70%
Number of Operators =	1
Hours per Day Required by Each Operator =	Varies†
Unit Labor Rate (including benefits) =	$50/hr

* Design run time of 400 bed volumes derived from pilot testing.
† 4 hours per day for ion exchange without brine treatment. 5 hours per day for ion exchange with brine treatment.

Engineering Assumptions for Cost Model

The ion exchange cost model was based on the ion exchange design capacity (flow rate, bed volumes, loading rates, etc.) and the spent brine water quality characteristics. There were multiple variables accounted for in the model. The major types of variables included design criteria, spent brine water quality (for discharge and/or treatment analysis), and cost information related to capital costs. Table 8.7 summarizes the engineering cost assumptions used in the engineering analysis. These values include both capital and operating costs, and are based on conceptual-level design assumptions. The level of detail of these analyses would indicate that a margin of error of -30 and +50 percent could be expected.

Table 8.8
Summary of assumptions used in cost model brine treatment options

Spent Brine Water Quality	***Value***
Perchlorate, mg/L =	3
Nitrate, mg/L =	1500
COD, mg/L =	3500
TSS, mg/L =	50
Capital Costs Associated With Biological or PNDM Treatment*	***Value***
Post Biological Filtration Unit =	\$15,000/40 gpm
Number of Pumps (Booster, Transfer, and Chemical) =	8
Pump Cost =	\$1000 per Hp
Biological Brine Treatment	***Value***
Number of Biological Reactor Tanks (16,500 gallon) =	5
Number of Equalization Tanks (3,000 gallons each) =	1
Biomass Seeing Requirements =	100 ml 1× year
Acetic Acid Requirements per Regeneration =	400 mg/L
Magnesium or Calcium Requirements =	800 mg/L 1 × month
Nitrogen Requirements =	Varies†
PNDM Brine Treatment	***Value***
Desired Ammonia Residual =	500 mg/L
Total Electricity Required for PNDM =	2,494,800 kw-hrs/year
Ammonia Reductant Cost =	\$500/ton
Catalyst Unit Cost in PNDM =	\$70/ton

* Membrane filtration units dependent upon reuse option (not included in brine line disposal options).
† Approximately 100 mg/L, but could possibly vary depending on operating conditions to maintain an anaerobic environment.

Table 8.8 summarizes the design parameters and costs associated with the brine treatment options evaluated in the pilot studies. Since the annual fee for use of a brine line is generally based on the chemical oxygen demand (COD) and the total suspended solids (TSS) concentration in untreated spent brine, these two parameters were included in the model and assumed at 3500 mg/L COD and 50 mg/L TSS based on brine water analysis collected during pilot testing.

Table 8.9
Perchlorate removal cost comparison

Type of ion exchange system	Brine management option	O&M cost ($/AF)	Capital costs* ($/AF)	Total cost† ($/AF)	Total cost† ($/1000 gals)
Conventional‡	Brine Line	317	85	402	1.23
LPVCWD ISEP	Brine Line	261	95	356	1.09
Conventional§	Biological	220	106	326	1.00
Conventional§	PDNM	291	172	463	1.42
LPVCWD ISEP+	PNDM	257	173	430	1.32

* Amortized for 20 years at 5% interest.
† Cost per acre-foot was rounded to the nearest $5 and cost per 1000 gallons was rounded to the nearest $0.05.
‡ Alternative 1 – No brine treatment with brine line discharge.
§ Alternative 2 – 6 Percent Brine treatment with brine reuse.

PERCHLORATE REMOVAL FACILITIES COMPARISON

The La Puente Valley County Water District is a perchlorate removal facility currently in operation. This facility utilizes a Calgon Carbon Corporation (Calgon) ISEP ion exchange system with brine line discharge. One years worth of O&M data, provided by the LPVCWD, was used as a cost comparison for the cost model discussed in this chapter. Calgon also provided additional data on estimated capital and O&M costs for an ISEP+ system (ion exchange plus PNDM) at LPVCWD.

Being both cost competitive and providing additional intangible benefits, Alternative 2 was selected for a full-scale cost comparison to existing perchlorate removal and/or destruction facilities. The results of the cost comparison are summarized in Table 8.9 and graphically displayed in Figure 8.3.

The full-scale treatment costs were relatively similar, with the least ($1.00) and most expensive ($1.42) options within 30 percent. The cost estimates were calculated using a conceptual level estimate with a margin of error of -30 and +50 percent. The conventional ion exchange plant with biological treatment and reuse was the least expensive at $1.10/1000 gallons ($360/AF) while conventional ion exchange with PNDM brine treatment was the most expensive at $1.43/1000 gallons ($466/AF). The following section details the variables that effect the LPVCWD costs.

La Puente Valley County Water District Ion Exchange Facility

The La Puente Valley County Water District (LPVCWD) currently uses a 2,500-gpm Calgon Carbon Corporation (Calgon) ISEP ion exchange system for perchlorate and nitrate removal. The ion exchange system is preceded by an air stripper for VOC removal and followed by an ultraviolet (UV) treatment system for NDMA removal. To maintain consistency when comparing full-scale ion exchange to MBfR technology, the capital and O&M costs associated with the air stripper and the UV system were removed from the total system costs.

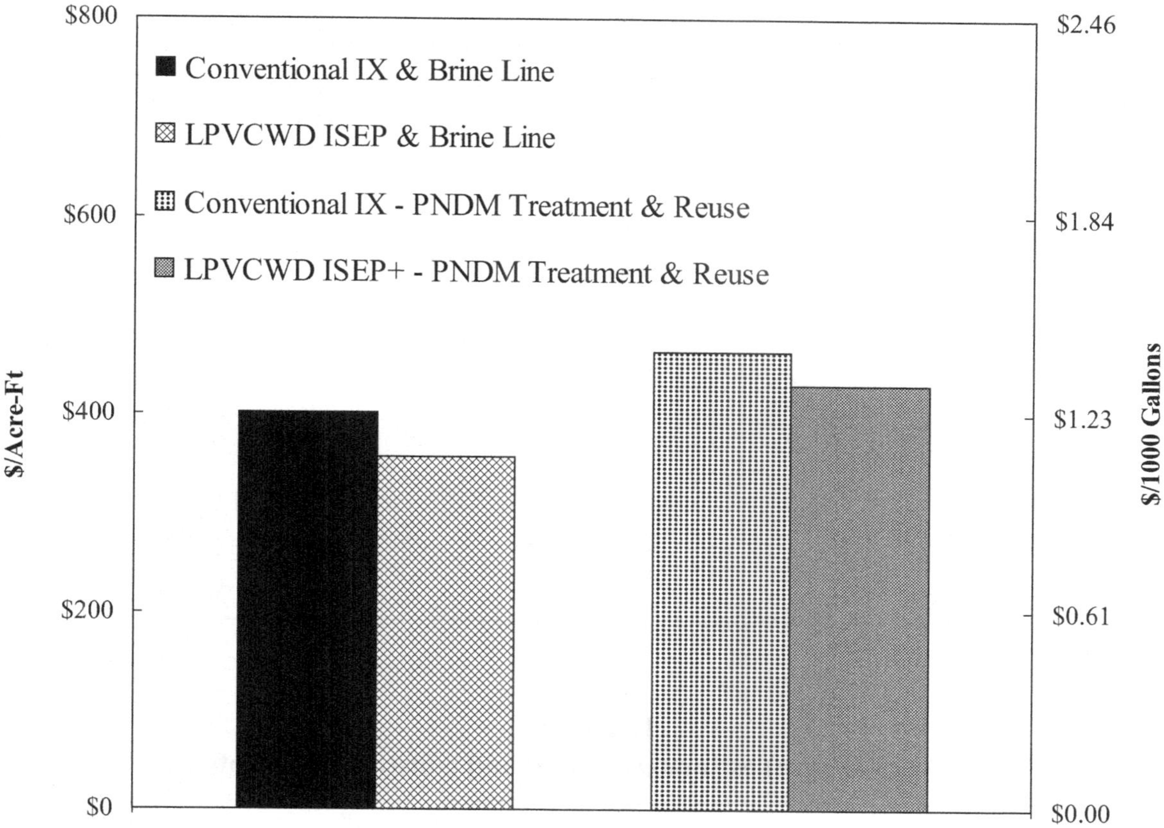

Figure 8.3 Full-scale cost comparison of ion exchange facilities

The ion exchange process produces a concentrated perchlorate- and nitrate-laden waste stream during the regeneration process. The perchlorate in the waste brine is not destroyed, but simply disposed of to a local county sanitation districts wastewater treatment facility through a dedicated brine discharge line. Future regulations may prevent the use of a brine line for discharging perchlorate-laden waste brine. With brine treatment being a possible future reality, the LPVCWD ion exchange system was evaluated based on both the current operating conditions and with estimated brine treatment costs. Costs associated with brine treatment were based on estimates provided by Calgon for their Perchlorate and Nitrate Destruction Module (PNDM). The PNDM is the brine treatment component of the Calgon ISEP+ system which treats and reuses spent brine from the ion-exchange regeneration process.

Annual O&M costs for the ion exchange system (no brine treatment) were provided by LPVCWD and Calgon. This data is presented in Table 8.10 and represents the costs associated with a production rate of 2500 gpm. The O&M costs do not include LPVCWD associated costs for water quality testing or other engineering related costs due to systems unrelated to the ion exchange system. The estimated capital costs for the 2500 gpm LPVCWD perchlorate removal system are presented in Table 8.11. Capital costs are presented for the ion exchange system, with and without brine treatment.

Table 8.10
LPVCWD (2500 gpm) estimated annual O&M costs

O&M parameter	No brine treatment	PNDM brine treatment
Salt	$406,000	$75,000
Maintenance	$166,000	$250,000
Other Consumables	$150,000	$227,000
Power	$252,000	$335,000
Utilities (water + gas)	$2,500	$58,000
Labor (+ overhead)	$45,000	$90,000
Brine Disposal	$33,000	$2,500
Total Cost per Year	$1,054,500	$1,037,500
Cost per Acre-Foot	$261/AF	$257/AF
Cost per 1000 Gallons	$0.80	$0.79

Table 8.11
LPVCWD (2500 gpm) estimated capital and construction related costs

Capital cost parameter	No brine treatment	Brine treatment
Brine Line Connection	$40,000	$0
ISEP	$2,200,000	$2,200,000
ISEP Installation	$770,000	$770,000
PNDM	$0	$2,000,000
PNDM Installation	$0	$700,000
Facility Structure	$240,000	$290,000
Construction Related Costs	$1,200,000	$2,100,000
EL&A Costs*	$355,000	$650,000
Total Cost to Construct	$4,805,000	$8,710,000
Amortized Capital Cost	$383,000/yr	$702,000/yr
Amortized Cost per Acre-Foot†	$95/AF	$174/AF
Cost per 1000 Gallons	$0.30	$0.53

* Engineering, Legal and Adminisrative Costs.
† 2500 gpm system at 5% interest for 20-years.

SUMMARY

Based on the engineering analysis provided in this chapter, both the biological and PNDM brine treatment options appear economically feasible. The following is a summary of the engineering analysis:

- Of the brine treatment alternatives evaluated the 6 percent biological brine treatment option was the least costly at $1.00/1000 gallons ($326/AF), closely followed by the 3 percent brine solution at $1.19/1000 gallons ($386/AF).
- The PNDM brine treatment option with brine line discharge of the treated brine was the most costly at $1.82/1000 gallons ($591/AF).
- Both the biological and PNDM brine treatment options were technically feasible alternatives (i.e. successful in destroying perchlorate and nitrate from the brine) and have intangible benefits compared discharging all of the contaminant-laden brine each cycle.

Additional items to consider regarding treating and recycling the brine are the minimizing the volume of waste discharged, minimizing impact to the environment through perchlorate and nitrate destruction, and preparing for possible future regulations that might require brine pretreatment prior to disposal. These conclusions assume that waste brine disposal is an option at the proposed treatment site.

CHAPTER 9
PROJECT CONCLUSIONS

Since the detection of perchlorate in some California groundwaters in early 1997, several studies have evaluated multiple treatment technologies for its removal. While several types of treatment systems have been designed to reduce perchlorate concentrations in contaminated water, biological treatment and ion exchange systems are among the technologies most frequently investigated. Though feasible, biological treatment of perchlorate-contaminated groundwater for potable use must first overcome significant public apprehension, even open hostility, before it can be applied in this manner. In the interim, the implementation of ion exchange, a proven and publicly accepted technology, is available to meet the existing treatment goals. The primary concern surrounding the use of ion exchange, however, is the specialized disposal and/or treatment of the perchlorate-laden saline waste stream generated as a by-product of the process.

In this research, the long-term performance of ion exchange using disparate alternatives for brine treatment and reuse was investigated. The first phase of the project (completed 2001), focused on the feasibility of ion exchange for perchlorate removal and the development of a salt-tolerant perchlorate-reducing culture.

Expanding on Phase I of the project, Phase II focused on;

- Long-Term Performance of Conventional Ion-Exchange for Perchlorate Removal. Two different types of resin (polyacrylic and polystyrene) were selected for further testing using contaminated groundwater from a currently operating drinking water well. Operational parameters were determined and baseline performance was established for both resin types.
- Effect of Heating on the Efficiency of Regeneration for Polystyrene Resins. The efficiency of non-heated regenerant (20°C) was compared to that of a regenerant heated up to 60°C in terms of effluent water quality and perchlorate leakage control.
- Brine Treatment and Reuse. Three different brine treatment alternatives were investigated: biological treatment; a physical/chemical high temperature and pressure catalytic process; and electrolytic perchlorate and nitrate reduction.

Based on previous bench-scale testing and recommendations from various resin manufactures, polyacrylic (Purolite A-850) and polystyrene (Bayer M500) resins were chosen for additional pilot testing.

Initial evaluations were performed with each resin to determine the appropriate operating conditions for the pilot plant by exhausting each resin to perchlorate breakthrough. Based on initial results, chromatographic peaking of nitrate concentrations in excess of the CaDHS operational goal of 8 mg-N/L for both resins limited the operation of a single vessel system in this water. Consequently, the run lengths selected were 425 BV and 625 BV for the polyacrylic and polystyrene resins, respectively. Counter-current regeneration of the exhausted resin using a 6 percent NaCl solution was performed until complete removal of perchlorate was achieved. Complete removal of nitrate, sulfate, and bicarbonate was achieved in 6 to 8 BV. Perchlorate was completely removed from the polyacrylic resin after 13 BV (62 lbs/ft^3). Based on this, a partial regeneration of 25 lbs/ft^3 was selected for the polyacrylic resin during baseline testing. For the polystyrene resin, counter-current regeneration was performed for over 140 BV (530 lbs/ft^3), at which point 800 μg/L perchlorate continued to be displaced from the resin.

Based on these results, a salt loading rate of 35 lbs/ft^3 (9 BV) was selected for the partial regeneration of the polystyrene resin during baseline testing.

Once operating conditions were established, both resins (polyacrylic and polystyrene) were continuously exhausted and regenerated to establish baseline performance. After 20 cycles, no change in performance, treated water quality or spent brine characteristics was detected. This performance served as a baseline to which three brine treatment and recycle process were validated.

The successful reuse of spent brine in the ion exchange process depends on the residual perchlorate level in the recycled brine following treatment. Maintaining a low level of perchlorate in the recycled brine is critical because of the resin's high affinity for perchlorate. Furthermore, perchlorate is a monovalent ion, which is not subject to selectivity reversal in the presence of high ionic strength NaCl brine used for regeneration. Owing to its inherent high affinity and lack of selectivity (affinity) reversal, a significant portion of any perchlorate remaining in the spent brine would re-adsorb (exchange) onto the resin during regeneration. Some of the re-adsorbed perchlorate will then exchange for chloride and leak into the column effluent, immediately from the start of the run subsequent cycle. Consequently, it necessary to reduce the perchlorate level in the brine before reuse.

Using the University of Houston EMCT-Windows model and by varying the amounts of perchlorate in a 6 percent brine for the regeneration of the polyacrylic resin, it was determined that the maximum allowable concentration of perchlorate in the treated 6 percent brine was approximately 0.8 mg/L (0.4 mg/L for a 3 percent brine). Manipulation of the model indicated that the equivalent fraction of perchlorate in the early leakage following regeneration would be the same as the equivalent fraction of perchlorate in the treated brine. Therefore, the perchlorate leakage could be quickly estimated as the product of the equivalent fraction perchlorate in the brine and the total concentration of ions in the feed water. The selected brine treatment goals were half of the theoretical limit to avoid any detectable leakage for each brine treatment system.

The biological brine treatment system was a 208 L (55 gal) closed reactor with multiple taps to introduce spent brine, draw-off treated brine, or sample water quality. The system was operated as a sequencing batch reactor (SBR). Spent brine (0.5 N / 3 percent NaCl) from the regeneration of the ion exchange resin, was first introduced to the nitrate and perchlorate degrading culture in the reactor and allowed to mix. Acetic acid was supplied to this culture as an electron donor for the reduction process, while the nitrate and perchlorate were biodegraded. The mixture was then allowed to settle under quiescent conditions for 1 hour and the clarified treated brine was drawn through two 5-15 μm cartridge filters to remove non-settleable bacteria. The chloride concentration of the filtered, treated brine was then adjusted before its reuse as the regenerant solution.

Results of the biological brine treatment system demonstrated that the process was able to consistently reduce the perchlorate and nitrate concentrations in the spent brine (0.5 N / 3 percent NaCl) to below the treatment goals ($ClO_4^- < 200$ μg/L and $NO_3^- < 0.5$ mg-N/L) for 20 (re)cycles. However, an increasing trend was observed in the concentration of nitrate ions in the finished water quality at the end of exhaustion. This clearly illustrated a slow decrease in the capacity of the resin for nitrate as various anions accumulated in the recycled brine and on the resin. This accumulation of other anions could be controlled by several methods:

- Periodically the recycled brine can be wasted and treatment continued with a fresh batch of brine,
- Periodically waste a small portion of the recycled brine (and subsequently amend it with virgin sweet brine) to limit the accumulation of these other anions,

- Further processing of the treated brine to remove some of these accumulating anions (i.e. passing the treated brine through an NF membrane).

Operationally, the number of times that brine may be recycled will depend on site-specific water quality, discharge permits, concerns regarding generation of hazardous waste (dependant upon raw water concentrations of heavy metals and radiological constituents), and the effect on the brine culture due to potential bioaccumulation of toxic contaminants.

The microecology study performed on a sample of pilot plant culture after 20 cycles of brine regeneration revealed that the culture was mixed. Pure cultures that could be developed on solid media were all facultative, aerobic, nitrate-reducing organisms. None were identified as known human pathogens. The one organism consistently seen in Gram stains, but resisted aerobic or anaerobic solid agar culture techniques, was a Gram negative vibrio and was suspected to be the salt-tolerant perchlorate reducing organism.

Separate mixed culture studies performed in the laboratory revealed that the preferential pH of the culture was between 8 and 9. Additional tests indicated that a balance between the monovalent and divalent cations was necessary for rapid ClO_4^- reduction (patent pending).

The chemical brine treatment system (PNDM – Perchlorate Nitrate Destruction Module) provided by Calgon Carbon Corporation (Pittsburgh, PA) employed a high-pressure and high-temperature catalytic process to reduce the nitrate and perchlorate in the spent brine. Following the PNDM, the treated brine was then ready for reuse without additional treatment. In addition to high pressure and temperature, the PNDM employed a chemical reductant (ammonia) dose based on the measured concentrations of nitrate and perchlorate in the spent brine.

Results demonstrated that the PNDM also consistently reduced the perchlorate and nitrate concentrations in the spent brine (1.0 N / 6 percent NaCl) to below detection limits (71 μg/L and 0.5 mg-N/L, respectively) during each cycle. Perchlorate and nitrate concentrations in the spent brine (prior to treatment) remained at approximately 3100 μg/L and 270 mg-N/L, respectively. The other major anions (sulfate and bicarbonate) were not removed by the PNDM and therefore consistently increased. After 30 (re)cycles, the results also demonstrated the increasing trend in the concentration of nitrate ions in the finished water quality at the end of exhaustion, similar to that which was observed with the biological treatment.

In addition to sulfate and bicarbonate, uranium, barium, fluoride, and silica accumulated in the brine after each cycle, as these constituents were not removed by the PNDM. While arsenic was adsorbed by the resin during exhaustion and displaced during regeneration, it was not detected in the brine after being processed by the PNDM. A loss of hardness through brine treatment was observed indicating that arsenic may have co-precipitated with $CaCO_3$. Similar to the biological brine treatment system, operationally, the number of time brine may be recycled will depend on site-specific water quality, discharge permits, and concerns of the generation of hazardous waste.

The goal of the electrochemical pilot testing was to evaluate if the pilot-scale electrochemical cell could reduce perchlorate in amended (perchlorate and nitrate) brine. During the trial, the cell was operated discontinuously for a total of 600 hours. Typically, during a process period of 16 hours, the nitrate levels were reduced from 20,000 mg/L to 4,000 mg/L. The trial showed that nitrate was reduced in the electrochemical cell, achieving power costs of 10 kWh/kg of nitrate reduced. Perchlorate reduction rates, however, were poor, as less than 10 percent of perchlorate was reduced even in the presence of low levels of nitrate. Pilot testing demonstrated that the electrochemical cell was effective at reducing nitrate in high chloride brines but it was concluded that further development and optimization would be required to achieve effective perchlorate removal rates.

An engineering analysis was performed to evaluate the cost of brine treatment and reuse. A cost model was developed to determine the capital and operation & maintenance costs associated with removing perchlorate from groundwater using a non-proprietary ion exchange system along with one of multiple forms of brine management. Results of the cost analysis revealed that biological and PNDM brine treatment options appear economically feasible. A conventional ion exchange system with brine discharge was estimated to cost \$1.23/1000 gallons (\$402/AF). Of the biological brine treatment alternatives evaluated, a 6 percent NaCl biological brine treatment option with reuse was the least costly at \$1.00/1000 gallons (\$326/AF), followed by a 3 percent NaCl biological system at \$1.19/1000 gallons (\$386/AF). The PNDM brine treatment option with reuse was estimated to cost \$1.42/1000 gallons (\$463/AF).

With increasing demands on water supplies and the continued discovery of perchlorate impaired waters, a proven and cost-effective means of treating perchlorate is needed. The implementation of ion exchange, a proven and publicly accepted technology, is available to meet the existing treatment goals. However, the primary concern surrounding the use of conventional ion exchange is the specialized disposal or treatment of the concentrated-perchlorate saline waste stream generated as a part of the process. In this research, it was found that innovative brine treatment strategies designed to treat spent regenerant brine generated from ion exchange processes can be used to effectively destroy perchlorate and nitrate facilitating brine reuse.

REFERENCES

APHA, AWWA, and WEF (American Public Health Association, American Water Works Association, and Water Environment Federation). 1998. *Standard Methods for the Examination of Water and Wastewater.* 20th ed. Washington, D.C.: APHA.

Baker, B.J., D.P. Moser, B.J. MacGregor, S. Fishbain, M. Wagner, N.K. Fry, B. Jackson, N. Speolstra, S. Loos, K. Takai, B.S. Lollar, J. Fredrickson, D. Balkwill, T.C. Onstott, C.F. Wimpee, and D.A. Stahl. 2003. Related Assemblages of Sulphate-Reducing Bacteria Associated With Ultradeep Gold Mines of South Africa and Deep Basalt Aquifers of Washington State. *Environ. Microbiol.*, 5(4):267–277.

Browner, C.M. 1999. Part II: Environmental Protection Agency. 40 CFT Parts 9, 141, and 142. Revisions to the Unregulated Contaminant Monitoring Regulation for Public Water Systems. Final Rule. *Fed. Reg.*, 64(180):50555–50620.

Cang, Y. 2001. Development of Cultures Capable of Reducing Perchlorate and Nitrate in Ion Exchange Regenerant Brine. Master's thesis. University of Houston, Houston, Texas.

Fonchy, E., A. Morin, N. Rodrigue, B. Müller, and P. Chalier. 1999. Effect of Growth Temperature on Hydrolytic and Esterifying Activities From *Pseudomonas fragi* CRDA 037 Grown on Whey. *Appl. Environ. Microbiol.*, 65:3114–3120.

Jackson, P.E., C. Weigert, C.A. Pohl, and C. Saini. 2000. Determination of Inorganic Anions in Environmental Waters With a Hydroxide-Selective Column. *Jour. Chromatogr. A*, 884:175–184.

Krieg, N.R., and J. Dobereiner. 1984. Genus Azospirillum. In *Bergey's Manual of Systematic Bacteriology.* Edited by N.R. Krieg et al. Williams & Wilkins. 94–103.

Lin, N.W., R. Zhang, R.Y. Zeng, and J. Zhao. 2002. Cold-Active Alkaline Protease From the Psychrophilic Bacterium Marinobacter Strain R2 Isolated From Antarctic Deep Sea Sediment [Online]. Available: <http://www.psc.edu/general/software/packages/genbank/genbank.html>.

Logan, B.E. 2001. Assessing the Outlook for Perchlorate Remediation. *Environ. Sci. Technol.*, 35:482A–487A.

Mayer, K. 2003. Memorandum—Transmittal of Perchlorate Maps and Table, updated April 2003. USEPA. May 14, 2003.

Palleroni, N.J. 1984. Family 1. Pseudomonadaceae. In *Bergey's Manual of Systematic Bacteriology.* Edited by N.R. Krieg et al. Williams & Wilkins. 172.

Perciasepe, R. 1998. Part III: Environmental Protection Agency. Announcement of the Drinking Water Contaminant Candidate List. Notice. *Fed. Reg.*, 63(40):10273–10287.

USEPA (U.S. Environmental Protection Agency). 2003. Ground Water & Drinking Water—Perchlorate. 23 January 2003 [Online]. Available: <http://www.epa.gov/safewater/ccl/perchlorate/perchlorate.html>. [cited 11 March 2003]

ABBREVIATIONS

BV	bed volume
°C	degrees Celsius
CaDHS	California Department of Health Services
CCL	Contaminant Candidate List
Cl^-	chloride
ClO_2^-	chlorite
ClO_3^-	chlorate
ClO_4^-	perchlorate
cm	centimeter
DO	dissolved (aqueous) oxygen
EBCT	empty-bed contact time
ft	feet
gpm	gallons per minute
HCl	hydrochloric acid
HRT	hydraulic residence time
IC	ion chromatograph
IX	ion exchange
L	liter
M	molar
MCL	maximum contaminant level
MDL	method detection limit
Mg	magnesium
mg/L	milligrams per liter
min	minute
mL	milliliter
mM	millimolar
MRL	method report limit
MWH	Montgomery Watson Harza
μg/L	micrograms per liter
μL	microliter
μS	microsiemens
NaCl	sodium chloride
NaOH	sodium hydroxide

ND	non detect
NH_4ClO_4	ammonium perchlorate
nm	nanometer
NO_2^-	nitrite
NO_3^-	nitrate
O_2	oxygen
OEHHA	Office of Environmental Health Hazard Assessment
PHG	public health goal
PLC	programmable logic controller
PNDM	Perchlorate and Nitrate Destruction Module
psi	pounds per square inch
PSU	power supply
PVC	polyvinyl chloride
QA	quality assurance
QAPP	quality assurance project plan
QC	quality control
RPD	relative percent deviation
SBR	sequencing batch reactor
UCMR	Unregulated Contaminants Monitoring Rule
USEPA	United States Environmental Protection Agency
VSS	volatile suspended solids

IWA Publishing's authorised EU representative for General Product
Safety Regulations is Diane D'Arras, 15 rue Duret, 75116 Paris,
France, e-mail: safety@iwap.co.uk.

Printed and bound by CPI Group (UK) Ltd, Croydon, CR0 4YY
17/09/2026
02218804-0014